ANNOTATED BIBLIOGRAPHY

OF EXPOSITORY WRITING

IN THE MATHEMATICAL SCIENCES

Matthew P. Gaffney
Univ. of Mass. at Boston

Lynn Arthur Steen
St. Olaf College

with the assistance of
Paul J. Campbell

The Mathematical Association of America

ISBN 0-88385-422-8

Copyright © 1976 by
The Mathematical Association of America

Printed in the United States of America

CONTENTS

Preface   v

Subject Classifications
1. General
   1.1 Survey   1
   1.2 History   4
   1.3 Biography   17
   1.4 Role of Mathematics   21
   1.5 Nature of Mathematics   25
   1.6 Pedagogy   32
   1.7 Miscellany   36

2. Foundations
   2.1 Set Theory   38
   2.2 Philosophy of Mathematics   43
   2.3 Intuitionism   47
   2.4 Nonstandard Analysis   49
   2.5 Decidability   52
   2.6 Logic   55

3. Algebra
   3.1 Combinatorics   60
   3.2 Graph Theory   63
   3.3 Number Theory   68
   3.4 Linear Algebra   75
   3.5 Group Theory   79
   3.6 Algebraic Structures   84

4. Analysis
    4.1 Elementary Calculus   91
    4.2 Advanced Calculus   95
    4.3 Real Analysis   99
    4.4 Complex Analysis   104
    4.5 Differential Equations   109
    4.6 Functional Analysis   115
    4.7 Special Topics   121

5. Geometry
    5.1 Classical Geometry   125
    5.2 Combinatorial Geometry   131
    5.3 Differential Geometry   133
    5.4 Algebraic Geometry   137
    5.5 General Topology   139
    5.6 Differential Topology   145
    5.7 Algebraic Topology   149
    5.8 Topological Groups   152

6. Statistics and Computing
    6.1 Probability   154
    6.2 Statistics   161
    6.3 Optimization   165
    6.4 Numerical Analysis   172
    6.5 Computer Science   177

7. Applications
    7.1 Physical Science   184
    7.2 Biological Science   196
    7.3 Behavioral Science   201
    7.4 Economic Science   204
    7.5 Information Science   208
    7.6 Miscellany   210

Index by Author   213

Preface

The purpose of this *Annotated Bibliography* is to facilitate access to the substantial but widely scattered expository literature of the mathematical sciences. It includes over 1100 articles (and a few books of general interest) that could be used to enrich undergraduate study of mathematics and that could be read by scientists and others who want accessible surveys of current topics in the mathematical sciences.

This bibliography is designed to be used in several different ways. Its primary purpose is to enable teachers to supplement their (or their students') background in standard undergraduate courses: almost every section in the bibliography could serve as a reading list for some undergraduate course. Others may wish to use the Index as a handy source of bibliographic data on articles they may have read long ago. Still others may find casual study of the bibliography rewarding in itself, for the annotations provide an unusual, diverse view of contemporary mathematics.

Our decision to focus the bibliography on expository articles (in English) at the undergraduate level was influenced by the availability of indices covering other parts of the mathematical literature. Textbooks and monographs are covered by the Telegraphic Reviews section of the *American Mathematical Monthly*, by the library magazine *Choice*, by the book review sections of many mathematical and scientific journals, and by the recently revised *Basic Library List* of the Committee on the Undergraduate Program in Mathematics (CUPM). The *Annotated Bibliography* complements the *Basic Library List* as a guide to the literature of undergraduate mathematics.

Research literature is reviewed regularly in *Mathematical Reviews* and in the various abstracting and reviewing journals of the applied mathematical sciences.

Although we did include some research-level survey articles, we view this area as the outer boundary of our objectives. Readers interested in research surveys should supplement our limited selection with items suggested by standard research indices.

The content of this *Bibliography* was necessarily limited by constraints of space and time. Although we had hoped to annotate every entry, it quickly became clear that this was an unrealistic expectation. So we included many items without (or with very brief) annotations, for it is surely better that they be listed than not. We must therefore emphasize that the length of an annotation, or its absence, is in no way intended as a judgment of the importance of the article. Because of these and other omissions, the present document is not a complete catalog. We believe, however, that it is a reasonably accurate reflection of the nature and quality of the expository literature in the mathematical sciences, and that it contains most of the very best of that genre.

The entries in this bibliography are arranged and cross-referenced by subject and then listed again by author to provide a detailed index. Within each subject area, articles are grouped according to level, and within each level they are arranged alphabetically.

The classification scheme set forth in the Contents follows generally the American Mathematical Society subject classification, modified in major ways to reflect the undergraduate curriculum and the extant expository literature. Wherever possible, common undergraduate courses are represented with appropriate sections. We make no philosophical or didactic claims for the structure of the bibliography: its major purpose is pragmatic- to enable a reader to locate as quickly as possible the kind of item he may be interested in.

Within each subject section we distinguish four levels of technical sophistication:

0. GENERAL. Requires, on average, no more technical background than that supplied by a good high school education.

1. ELEMENTARY. Requires some familiarity with work covered in the first two years of college mathematics, especially calculus and linear algebra.

2. ADVANCED. Requires knowledge of some topics that are ordinarily included in the upperclass undergraduate mathematics curriculum.

3. RESEARCH. Requires familiarity with graduate level work in the appropriate area.

We know from countless examples that many articles simply do not fit properly into any of these four categories. Nonetheless, the assigned levels do provide some clue, albeit imperfect, to the general difficulty of the article.

The subject and level of each article is coded into a three digit symbol, e.g., [3.2.1], that is used throughout the bibliography as a standard means of reference. The first digit is the number of the chapter, the second digit the number of the section, and the third digit the level. Thus an article labelled [3.2.1] would be located in the Elementary subsection of the Combinatorics section of the Algebra chapter.

Each article is listed with full bibliographic data and annotation (if present) in the section referred to by its code. Bibliographic data (without annotation) is repeated, where appropriate, in various Related References subsections that conclude each subject section, and again in the complete Index by Author. Wherever an item appears as a secondary entry it is accompanied by its three digit code, which indicates the location of the primary reference.

The articles in this bibliography together with the books listed in the *Basic Library List* provide extensive coverage of the undergraduate literature in mathematics. Other indices provide access to more specialized areas:

Dick, Elie M., *Current Information Sources in Mathematics*, Libraries Unlimited, 1973.

Fang, J., *A Guide to the Literature of Mathematics Today*, Paideia Pr., 1972.

Invited Addresses, 1965-1974, *Bull. Amer. Math. Soc.*, 81 (1975) General Index, pp. 38-42.

May, Kenneth O., *Bibliography and Research Manual of the History of Mathematics*, U. Toronto Pr., 1973.

Parke, M., *Guide to the Literature of Mathematics and Physics*, Dover, 1958.

Pemberton, John E., *How to Find Out in Mathematics*, 2nd rev. ed., Pergamon, 1970.

Pritchard, A., *A Guide to Computer Literature*, Archon Books, 1969.

Schaaf, William L., *A Bibliography of Recreational Mathematics*, Vols. 1-3, NCTM, 1970.

Many of the articles included in this bibliography appear in volumes of essays that contain either original or reprinted papers. Such volumes form a fine basic library of mathematical exposition. Here is a list of some common and popular volumes:

Aleksandrov, A.D., Kolmogorov, A.N. and Lavrent'ev, M.A., *Mathematics: Its Content, Methods, and Meaning*, Vols. I-III, MIT Pr., 1963.

Apostol, Tom M., et al., *Selected Papers on Calculus*, M.A.A., 1968.

Benacerraf, Paul and Putnam, Hilary, *Philosophy of Mathematics: Selected Readings*, Prentice-Hall, 1964.

Behnke, H., et al., *Fundamentals of Mathematics*, Vols. I-III, MIT Pr., 1974.

Feigenbaum, Edward A. and Feldman, Julian, *Computers and Thought*, McGraw-Hill, 1963.

Fenickel, Robert R. and Weizenbaum, Joseph, *Computers and Computation: Readings from Scientific American*, Freeman, 1971.

Kline, Morris, *Mathematics in the Modern World: Readings from Scientific American*, Freeman, 1968.

LeLionnais, F., *Great Currents of Mathematical Thought*, Vols. I-II, Dover, 1971.

May, Kenneth O., *Lectures on Calculus*, Holden-Day, 1967.

Newman, James R., *The World of Mathematics*, Vols. 1-4, Simon and Schuster, 1956.

Saaty, Thomas L. and Weyl, F. Joachim, *The Spirit and the Uses of the Mathematical Sciences*, McGraw-Hill, 1969.

Saaty, Thomas L., *Lectures on Modern Mathematics*, Vols. I-III, Wiley, 1963-65.

Schaaf, William L., *Our Mathematical Heritage*, Collier, 1963.

Studies in Mathematics, Vols. 1-12, M.A.A., 1962-1976.

Tanur, Judith M., et al., *Statistics: A Guide to the Unknown*, Holden-Day, 1972.

The Mathematical Sciences, A Collection of Essays, MIT Pr., 1969.

Other articles--some excellent, some worthless--can be found in the various encyclopedias. We have included in this volume many articles found in the most recent editions of the two major general encyclopedias--*Britannica III* and the 1976 edition of *Encyclopedia Americana*--as well as in the *McGraw-Hill Encyclopedia of Science and Technology*. Earlier editions of the *Britannica* contain several excellent articles that are not included in *Britannica III* or in this *Bibliography*. We have also omitted references to the many biographical articles in these encyclopedias and in such specialized sources as Scribner's *Dictionary of Scientific Biography*. Such articles, which are easy to locate because of the alphabetical organization of encyclopedias and dictionaries, provide a valuable supplement to our references on history and biography.

We expect that each reader may find sections in which our selection is at variance with his own favorite list. If the *Bibliography* turns out to be as useful to the mathematical community as we hope it will, we will quite likely update and revise it at some future time. We therefore welcome suggestions for how it might be modified, including suggestions for additions, deletions, annotations, and organization. (Since the editorial files for the bibliography are maintained by Lynn Steen in the M.A.A. editorial office at St. Olaf College, Northfield, Minn., 55057, material for this possible revision should be sent to him.)

    The lineage of this bibliography extends in two directions--to a bibliography prepared by Matthew Gaffney for the Conference Board of the Mathematical Sciences under a grant from the Office of Naval Research, and to a computer-based bibliography of undergraduate level exposition prepared by Lynn Steen for the Mathematics Department of St. Olaf College. Following completion of Gaffney's bibliography, the Conference Board asked the Publications Committee of the Mathematical Association of America to continue development of the bibliography for eventual publication. The present document is a direct consequence of that request.

    Many people have contributed to this bibliography by suggesting items for inclusion, by preparing annotations, and by providing constructive criticism at various stages in its development. We wish especially to acknowledge the criticism and assistance of Garrett Birkhoff, Truman Botts, Thomas Hawkins and Reuben Hersh who made major contributions at different stages in the evolution of this work. Paul J. Campbell provided extraordinary assistance by reading in detail several earlier versions of the bibliography, and by discovering scores of references in relatively obscure sources of which we would not otherwise have been aware. Jennifer Galovich copyedited and proofread the entire manuscript, assiduously checking and cross-checking innumerable details. All typing for the bibliography, including two preliminary versions and this final copy, was done with extraordinary proficiency by Mary Kay Peterson. We are indebted to these people for their diligence and continuing good humor.

June 1, 1976
                                    Matthew P. Gaffney
                                    Boston, Massachusetts

                                    Lynn Arthur Steen
                                    Northfield, Minnesota

1.1 Survey

1.1.0 GENERAL

Boehm, George A.W., <u>The New World of Mathematics</u>, The Dial Press, 1959.

   This book reprints two 1958 <u>Fortune</u> articles on mathematics and a 1959 article on computers. The list of mathematicians that the author consulted reads like a Who's Who of U.S. mathematics, and the articles might be viewed as a collective judgment of what should go into expository articles for the layman.

Dantzig, Tobias, <u>Number, The Language of Science</u>, Doubleday Anchor, 1956.

Eves, Howard W., *Mathematics*, <u>Encyclopedia Americana</u>, 1976, V. 18, pp. 431-434.

Kline, Morris, <u>Mathematics in Western Culture</u>, Oxford U. Pr., 1953.

Kramer, Edna E., <u>The Nature and Growth of Modern Mathematics</u>, Hawthorn, 1970; Fawcett, 1973.

Pedoe, Daniel, <u>The Gentle Art of Mathematics</u>, Macmillan, 1958.

Singh, Jagjit, <u>Great Ideas of Modern Mathematics: Their Nature and Use</u>, Dover, 1959.

Steinhaus, Hugo, <u>Mathematical Snapshots</u>, Oxford U. Pr., 1969.

   A highly regarded series of mathematical vignettes that convey the nature of various mathematical notions with a minimum use of formulas and a maximum use of diagrams.

Stewart, Ian, <u>Concepts of Modern Mathematics</u>, Penguin, 1975.

Stone, Marshall H., *The revolution in mathematics*, <u>Liberal Education</u>, 47 (1961) 304-327; also in <u>Amer. Math. Monthly</u>, 68 (1961) 715-734.

   Excellent general survey of the explosive growth of contemporary mathematics and the implications of this growth for college mathematics education.

Temple, G., *The growth of mathematics*, <u>Math. Gazette</u>, 41 (1957) 161-168.

The state of the mathematical sciences, in The Mathematical Sciences--A Report, National Academy of Sciences, 1968, pp. 45-116.

  Comprehensive survey of pure and applied mathematics written for those who make public policy.

Waismann, Friedrich, Introduction to Mathematical Thinking, Harper & Brothers, 1959.

Whitehead, Alfred North, An Introduction to Mathematics, Oxford U. Pr., 1958.

1.1.1 ELEMENTARY

Bochner, Salomon, *Mathematics*, McGraw-Hill Encyclopedia of Science and Technology, 1960, V. 8, pp. 175-180.

Courant, Richard and Robbins, Herbert, What is Mathematics?, Oxford U. Pr., 1941.

  One of the most famous and successful of all the expository books on selected mathematical topics.

Rademacher, Hans and Toeplitz, Otto, The Enjoyment of Mathematics: Selections from Mathematics for the Amateur, Princeton U. Pr., 1957.

  28 independent lectures--many on number theory--aimed at the interested mathematical amateur. Sample titles: On Waring's Problem, The Four-Color Problem, The Regular Polyhedrons, The Figure of Greatest Area with a Given Perimeter.

Tietze, Heinrich, Famous Problems of Mathematics, Graylock Pr., 1965.

  Readable accounts, with historical asides, of famous problems involving number theory, geodesics, higher dimensional spaces, map coloring, space curvature, and other topics.

Weil, André, *The future of mathematics*, Amer. Math. Monthly, 57 (1950) 295-306; also in F. LeLionnais, Great Currents of Mathematical Thought, V. 1, Dover, 1971, pp. 321-336.

  A general survey of the nature of mathematics (now somewhat dated) that is convincingly illustrated by frequent allusions to problems at the then-current research frontier.

Weyl, Hermann, *A half-century of mathematics*, Amer. Math. Monthly, 58 (1951) 523-553.

  The author reviews an era in which he himself was a major figure to "explain the most outstanding mathematical notions devised, and list some of the more important problems solved, in this period." Some of the pages read as though lifted from a liberal arts mathematics text (or is it the other way around?), but that just underlines the author's effort to make the review accessible to all interested parties. The article makes free use of physical as well as mathematical illustrations, as in "strangely enough [the ergodic hypothesis] was proved shortly after the transition from classical to quantum mechanics had rendered the hypotheses almost valueless, and it was proved by making use of the mathematical apparatus of quantum physics."

1.1.2 ADVANCED

Dieudonné, Jean A., *Recent developments in mathematics*, Amer. Math. Monthly, 71 (1964) 239-248.

A wide-ranging discussion, expressing the opinion that a sizable part of the mathematics done after 1945 has consisted in radically new departures heralding another era in mathematics. Although it contains few formulas, the essay's numerous brief references to then-current conjectures and problems makes for rather strenuous reading.

Kac, Mark and Ulam, Stanislaw M., Mathematics and Logic: Retrospect and Prospects, Frederick A. Praeger, 1968; The New American Library, 1969.

A beautifully written exposition of mathematics containing a bewildering array of topics. The authors--two masters of pure and applied mathematics--strive to develop a sense of interconnection between them.

1.2 HISTORY

1.2.0 GENERAL

Crowe, Michael J., *Ten "laws" concerning patterns of change in the history of mathematics*, Historia Math., 2 (1975) 161-166.

David, F.N., Games, Gods and Gambling, Hafner, 1962.

  A history of the theory of probability.

Dehn, Max, *Mathematics*, Amer. Math. Monthly; *400 B.C.-300 B.C.*, 50 (1943) 411-414; *300 B.C.-200 B.C.*, 51 (1944) 25-31; *200 B.C.-600 A.D.*, 51 (1944) 149-157.

Fisher, Charles S., *Some social characteristics of mathematicians and their work*, Amer. J. Sociology, 78 (1973) 1094-1118.

  "The social characteristics of three generations of mathematicians who have been engaged in...the Poincaré conjecture, are examined."

Greenblatt, M.H., *The "legal" value of π, and some related mathematical anomalies*, Amer. Scientist, 53 (1965) 427A-434A.

  History of the attempt of the Indiana State Legislature to set the value of pi at 3.

Knuth, Donald E., *Ancient Babylonian algorithms*, Comm. Assoc. Comp. Mach., 15 (1972) 671-677.

  Excellent piece of historical research, combined with a computer scientist's perspective.

Kolata, Gina Bari, *Mathematical problems: a committee to replace Hilbert*, Science, 185 (1974) 430.

  Very brief report on current efforts to produce a new Hilbert-like list of unsolved problems.

Le Corbeiller, P., *The curvature of space*, Scientific American, 191 (November 1954) 80-86, 124; also in M. Kline, Mathematics in the Modern World, Freeman, 1968, pp. 128-133, 397.

  An excellent popular account of the origins of differential geometry and of the notion of curvature of space--ordinary space or n-dimensional space. It explains some of the notions very clearly and has interesting historical remarks, particularly concerning Riemann's contributions.

LeVeque, William J., et al., *History of mathematics*, Encyclopaedia Britannica, 15th ed., 1974, Macropaedia, V. 11, pp. 639-670.

Mahoney, Michael S., *Another look at Greek geometrical analysis*, Arch. Hist. Exact Sci., 5 (1968) 318-348.
   An interesting discussion, illuminated by many examples, of what the Greeks meant by analysis. Provides helpful background material for an appreciation of the "mathematical revolution" of the 17th century, which was based upon a transformation of the Greek conception of analysis.

Meserve, Bruce E., *Number systems and notation*, Encyclopedia Americana, 1976, V. 20, pp. 536f-536j.

Morgan, Bryan, Men and Discoveries in Mathematics, Transatlantic Arts, 1972.

Neményi, P.F., *The main concepts and ideas of fluid dynamics in their historical development*, Arch. Hist. Exact Sci., 2 (1962) 52-86.
   Descriptive, non-technical, no equations.

Rosen, Saul, *Electronic computers: a historical survey*, Computing Surveys, 1 (1969) 7-36.
   A general overview of the development of computers, 1943-1969, with bibliography.

Seidenberg, A., *On the area of a semi-circle*, Arch. Hist. Exact Sci., 9 (1972) 171-211.
   The main thesis is that the geometries of Greece, Babylonia, Egypt, India and China are a derivative of a system of ritual practices as disclosed in the Sulvasútras, an Indian sacred work on altar constructions. Gives various supporting examples together with several other startling theses. The author is one of the few western mathematicians to explore Indian mathematics in depth.

Sheynin, O.B., *On the prehistory of the theory of probability*, Arch. Hist. Exact Sci., 12 (1974) 97-141.

Wilder, Raymond L., Evolution of Mathematical Concepts: An Elementary Study, John Wiley, 1968.
   An analysis of mathematics--"a cultural entity subject to influences that have directed and controlled its evolution"--as "one of the most important cultural components of every modern society."

Wilder, Raymond L., *Hereditary stress as a cultural force in mathematics*, Historia Math., 1 (1974) 29-46.

> An investigation of mathematics from the unusual perspective of its being a cultural system. Includes an analysis of the components of "growth pressure" in mathematics as it arises as an internal influence.

Wilson, Curtis, *How did Kepler discover his first two laws?*, Scientific American, 226 (March 1972) 92-106, 126.

> "It is generally assumed that he did so by calculating the distances between a planet and the sun and then perceiving that the distances fitted into an ellipse. It is more likely that the ellipse came first."

1.2.1 ELEMENTARY

Aaboe, Asger, Episodes From the Early History of Mathematics, New Math. Libr., No. 13, Random House, 1964; M.A.A., 1975.

> "Babylonian mathematics [as] recovered from cuneiform texts only during the last half century; Euclid's construction of the regular pentagon from his Elements; three small samples of Archimedes' mathematics: his trisection of an angle, his construction of the regular heptagon, and his discovery of the volume and surface of a sphere; and, lastly, Greek trigonometry as it is presented by Ptolemy in his Almagest."

Baron, Margaret E., The Origins of the Infinitesimal Calculus, Pergamon, 1969.

Bell, Eric Temple, *Newton after three centuries*, Amer. Math. Monthly, 49 (1942) 553-575.

> Detailed account of his work, with extensive bibliography.

Bell, Eric Temple, The Development of Mathematics, McGraw-Hill, 1945.

> A well-known book, often referred to. While it tends to oversimplify, it makes for lively reading.

Bergmann, Peter G., *Fifty years of relativity*, Science, 123 (1956) 487-494.

Bochner, Salomon, *Mathematical reflections*, Amer. Math. Monthly, 81 (1974) 827-852.

> Part I, *Purely Mathematical Americana*, contains American mathematical consequences of such miscellany as that it was not Euclid, Fermat, or Euler who first stated that an integer is uniquely

decomposable as a product of primes, but Gauss; how Carathéodory came to abandon his long search for a definitive proof of the Jordan curve theorem; and the origins of linear associative algebra. Part II, *Charles Sanders Peirce*, can be viewed as a separate article. "His was a real American tragedy. A great philosopher, and an even greater failure." For contrary views see rebuttal notes by Carolyn Eisele and Max H. Fisch in <u>Amer. Math. Monthly</u>, 82 (1975) 477-480.

Bochner, Salomon, *The role of mathematics in the rise of mechanics*, Amer. Scientist, 50 (1962) 294-311.

A survey up to the beginning of the nineteenth century.

Bos, H.J.M., *Differentials, higher-order differentials and the derivative in the Leibnizian calculus*, <u>Arch. Hist. Exact Sci.</u>, 14 (1974) 1-90.

A lucid study of the conceptual changes the calculus underwent in the period from Leibniz to Euler. Special attention is given to the reasons for the emergence of functions and derivatives, rather than differentials, as the fundamental concepts.

Chittenden, J. Brace, *Quadrature of the circle*, <u>Encyclopedia Americana</u>, 1976, V. 23, pp. 52-53.

Coolidge, J.L., *The lengths of curves*, <u>Amer. Math. Monthly</u>, 60 (1953) 89-93.

Brief history of attempt to define curve length.

Coxeter, H.S. MacDonald, *The space-time continuum*, <u>Historia Math.</u>, 2 (1975) 289-298.

Davis, Chandler and Pogorzelski, H.A., *Contemporary mathematical notation*, <u>McGraw-Hill Encyclopedia of Science and Technology</u>, 1960, V. 8, pp. 172-174.

Desanti, Jean T., *From Cauchy to Riemann, or the birth of the theory of real functions*, in F. LeLionnais, <u>Great Currents of Mathematical Thought</u>, V. 1, Dover, 1971, pp. 181-190.

Concise, relatively easy survey of nineteenth century continuity and integrability from a historical perspective.

Drake, Stillman, *Galileo's discovery of the law of free fall*, <u>Scientific American</u>, 228 (May 1973) 84-92, 120.

"It has been thought that [Galileo] erroneously assumed that the velocities of a falling body were proportional to distances. A new manuscript shows that he treated them correctly as being proportional to time."

Drake, Stillman, *Mathematics and discovery in Galileo's physics*, Historia Math., 1 (1974) 129-150.
> "Galileo's steps in the discovery of the law of free fall and its application to inclined planes are retraced from one of his letters and some manuscript notes."

Fang, J., *Hilbert's problems*, Philosophia Mathematica, 6 (1969) 38-53.

Fisher, Charles S., *The death of a mathematical theory: a study in the sociology of knowledge*, Arch. Hist. Exact Sci., 3 (1966) 137-159.
> On invariant theory. Very unusual essay, non-technical in nature.

Gleason, Andrew M., *Evolution of an active mathematical theory*, Science, 145 (1964) 451-457; also appeared as *The evolution of differential topology*, in The Mathematical Sciences--A Collection of Essays for COSRIMS, M.I.T., 1969, pp. 176-189.
> The author discusses topology, using as his main theme its relevance to the n-body problem of the Newtonian theory of planetary motion, in which a series of integrations of differential equations leads to surfaces in state-space. This makes an excellent expository vehicle for various aspects of topology.

Kiernan, B. Melvin, *The development of Galois theory from Lagrange to Artin*, Arch. Hist. Exact Sci., 8 (1971) 40-154.

Kline, Morris, *Geometry: history and development*, Encyclopedia Americana, 1976, V. 12, pp. 471-478.

Kline, Morris, Mathematical Thought from Ancient to Modern Times, Oxford U. Pr., 1972.
> A monumental work that traces the roots and development of the major branches of mathematics with special attention to its interplay with other fields.

Koppelman, Elaine, *The calculus of operations and the rise of abstract algebra*, Arch. Hist. Exact Sci., 8 (1971) 155-242.

Langer, R.E., *Fourier's series--the genesis and evolution of a theory*, Amer. Math. Monthly, 54 (1947) Suppl. pp. 1-86.
> Excellent exposition with an especially useful commentary on relation of vibrating string and heat flow to Fourier series. "In its modern form the theory of Fourier's series and its application to problems of physics admit of presentation in a direct and

logical manner that is, on the whole, strikingly economical in design. ...[O]f erstwhile possible deficiencies, no trace is left revealed. Well developed mathematical theories are prone to seem like that, and in the deceptiveness of this there is weakness as well as strength. ...[W]hen, as in the present instance, the ingenuity and the technical exploits which gave the impetus and direction were those of such masters as the Bernoullis, Euler, D'Alembert, Lagrange and Fourier, it need hardly be feared that a review of the course of the developments will prove to be an unrewarding venture."

Mahoney, Michael S., *Fermat's mathematics: proofs and conjectures*, Science, 178 (1972) 30-36.

A survey of Fermat's working habits as a mathematician to analyse his unwritten proof of the "last theorem." Concludes that Fermat extrapolated without proof that the method he had used to prove the theorem for the third and fourth powers would serve in general.

Pierpont, James, *Mathematical rigor, past and present*, Bull. Amer. Math. Soc., 34 (1928) 23-53.

Review of 17th and 18th century reasoning in calculus followed by an introduction to logicism, formalism and intuitionism.

Rosenthal, Arthur, *The history of calculus*, Amer. Math. Monthly, 58 (1951) 75-86.

From Archimedes to Newton Leibniz. Good bibliography.

Shenton, Walter F., *Mathematical signs and symbols*, Encyclopedia Americana, 1976, V. 18, pp. 426-428.

Taylor, Angus E., *The differential: nineteenth and twentieth century developments*, Arch. Hist. Exact Sci., 12 (1974) 355-383.

1.2.2 ADVANCED

Ayoub, Raymond, *Euler and the zeta function*, Amer. Math. Monthly, 81 (1974) 1067-1086; Addendum, 82 (1975) 737.

After a brief synopsis of Euler's life, the author describes various aspects of his work on the zeta function, including the methods by which in 1734 Euler finally obtained the closed form $\zeta(2) = \pi^2/6$ (as well as $\zeta(2n)$ up to $n = 6$). He also describes Euler's anticipation of Riemann's functional equation for the zeta function. Particularly valuable since virtually none of Euler's original memoirs have been translated into English. Winner of the Lester R. Ford award for expository writing.

Bell, Eric Temple, *Gauss and the early development of algebraic numbers*, National Math. Magazine, 18 (1944) 188-204, 219-233.

Birkhoff, Garrett, *Current trends in algebra*, Amer. Math. Monthly, 80 (1973) 760-782; Correction, 81 (1974) 746.

> Part I gives "...a brief résumé of the development of algebra as we know it today, over the past several centuries." In 16 concise but informative sections various aspects of algebra and symbolic logic are discussed, culminating in "The Reign of Modern Algebra, 1930-1970." Part II, on the influence of computers begins: "Already in the 1940's a new revolution was brewing, whose ultimate implications for mathematics are unpredictable." The author's thesis is clearly that Part II describes where the action is today. Winner of the Lester R. Ford award for expository writing.

Bochner, Salomon, *The rise of functions*, Rice Univ. Studies, 56:2 (1970) 3-21.

> General history of functions with emphasis on Fourier and complex analysis. Easy reading.

Dauben, Joseph W., *Denumerability and dimension: the origins of Georg Cantor's theory of sets*, Rete, 2 (1974) 105-133.

Dauben, Joseph W., *The invariance of dimension: problems in the early development of set theory and topology*, Historia Math., 2 (1975) 273-288.

Dauben, Joseph W., *The trigonometric background to Georg Cantor's theory of sets*, Arch. Hist. Exact Sci., 7 (1971) 181-216.

Davis, Philip J., *Leonard Euler's integral: a historical profile of the gamma function*, Amer. Math. Monthly, 66 (1959) 849-869.

> Very lively history mixed with aesthetics and opinion. A beautiful account of the gamma function and related areas of classical analysis. Winner of the Chauvenet Prize for expository writing in mathematics.

Goldstein, Marie, *The historical development of group theoretical ideas in connection with Euclid's axiom of congruence*, Notre Dame J. of Formal Logic, 13 (1972) 331-349.

> Example of group-theoretic ideas from editors of and commentaries on Euclid through the ages, aiming to show that "group theoretic ideas have been present in the minds of men since ancient times."

Grattan-Guinness, Ivor, The Development of the Foundations of Mathematical Analysis from Euler to Riemann, M.I.T. Pr., 1970.

Well-written and provocative with some controversial interpretations.

Hawkins, Thomas, *Cauchy and the spectral theory of matrices*, Historia Math., 2 (1975) 1-29.

Cauchy provided in 1829 the first general proof that the eigenvalues of a symmetric matrix are real, and his paper initiated further developments that resulted in a substantial spectral theory of matrices by the 1870's. The essay considers Cauchy's work and evaluates its historical significance.

Hawkins, Thomas, Lebesgue's Theory of Integration: Its Origins and Development, U. of Wisc. Pr., 1970; Chelsea, 1975.

A historical monograph based on the author's Ph.D. dissertation in mathematics and the history of science.

Hawkins, Thomas, *The theory of matrices in the 19th century*, in Proc. Inter. Cong. Math. (1974), V. 2, Canad. Cong. Math., 1975, pp. 561-570.

The author outlines the development of matrix theory in the 19th century, starting with Gauss's use of linear substitutions in his 1801 study of the arithmetical theory of quadratic forms. He emphasizes that matrix theory derived from a variety of sources, and pays particular attention to the development of the spectral theory of matrices and to its relationship to 18th century work on systems of differential equations. The author debunks what he dubs the "Cayley-as-Founder" view of matrix theory history.

Lyusternik, L.A., *The early years of the Moscow mathematical school*, Russian Math. Surveys, 22:1 (1967) 133-157; 22:2 (1967) 171-211; 22:4 (1967) 55-91; 25:4 (1970) 167-174.

Maistrov, L.E., Probability Theory: A Historical Sketch, Academic Pr., 1974.

Manheim, Jerome H., The Genesis of Point Set Topology, Pergamon, 1964.

McHugh, James A.M., *An historical survey of ordinary linear differential equations with a large parameter and turning points*, Arch. Hist. Exact Sci., 7 (1971) 277-324.

Mostowski, Andrezej, Thirty Years of Foundational Studies, Barnes & Noble, 1966.

Nový, Luboš, The Origins of Modern Algebra, Noordhoff, 1974.

> Traces the important changes that occurred in algebraic research in the 19th century and formed a necessary precondition for the rise of modern algebra.

Pierpont, James, *The history of mathematics in the nineteenth century*, Bull. Amer. Math. Soc., 11 (1904) 136-159.

Weyl, Hermann, *Relativity theory as a stimulus in mathematical research*, Proc. Amer. Phil. Soc., 93 (1949) 535-541.

1.2.3 RESEARCH

Bernkopf, Michael, *A history of infinite matrices*, Arch. Hist. Exact Sci., 4 (1968) 308-358.

> From birth in 1890's to death in 1930 when von Neumann introduced operator theory.

Bernkopf, Michael, *The development of function spaces with particular reference to their origins in integral equation theory*, Arch. Hist. Exact Sci., 3 (1966) 1-96.

> Detailed survey from Hilbert (1904) to Banach (1930). Largely an organized English summary of the major papers of Hilbert.

Coppel, W.A., *J.B. Fourier--On the occasion of his two hundredth birthday*, Amer. Math. Monthly, 76 (1969) 468-483.

> "Fourier's work on the conduction of heat has stimulated the most diverse developments in pure mathematics... [We] trace these developments in outline." The result is a thumbnail sketch of the mathematical descendents of Fourier's series, including the Riemann and Lebesgue integrals, Bohr's study of almost periodic functions and Fourier analysis on groups. An excellent survey which demonstrates the intimate relationship between group theory and modern analysis. Winner of the Lester R. Ford award for expository writing.

Dieudonné, Jean A., *The historical development of algebraic geometry*, Amer. Math. Monthly, 79 (1972) 827-866.

> An interesting condensed account of the historical development of algebraic geometry best suited to those with some prior knowledge of the subject. Winner of the Lester R. Ford award for expository writing.

Hawkins, Thomas, *Hypercomplex numbers, Lie groups and the creation of group representation theory*, Arch. Hist. Exact Sci., 8 (1972) 243-287.

Hawkins, Thomas, *New light on Frobenius' creation of the theory of group characters*, Arch. Hist. Exact Sci., 12 (1974) 217-243.

Based on recently discovered manuscripts.

Hawkins, Thomas, *The origins of the theory of group characters*, Arch. Hist. Exact Sci., 7 (1971) 142-170.

Lefschetz, Solomon, *The early development of algebraic geometry*, Amer. Math. Monthly, 76 (1969) 451-460.

Comments on various 19th century developments in algebraic geometry, as seen by one of the great masters of the subject.

Monna, A.F., Functional Analysis in Historical Perspective, Halsted Pr., 1973.

van der Waerden, B.L., *The foundation of algebraic geometry from Severi to André Weil*, Arch. Hist. Exact Sci., 7 (1971) 171-180.

RELATED REFERENCES

Agazzi, Evandro, *The rise of the foundational research in mathematics*, Synthese, 27 (1974) 7-26. [2.2.0]

Aleksandrov, P.S., *Poincaré and topology*, Russian Math. Surveys, 27:1 (1972) 157-168. [5.5.2]

Archibald, R.C., *The first translation of Euclid's Elements into English and its source*, Amer. Math. Monthly, 57 (1950) 443-452. [5.1.0]

Bhat, U. Narayan, *Sixty years of queueing theory*, Management Science, 15:6 (1969) B280-B294. [6.1.2]

Brush, Stephen G., *Foundations of statistical mechanics 1845-1915*, Arch. Hist. Exact Sci., 4 (1967) 145-183. [7.1.2]

Chern, S.S., *Differential geometry: its past and its future*, in Actes Cong. Inter. Math. (1970), V. 1, Gauthier-Villars, 1971, pp. 41-53. [5.3.3]

Coolidge, J.L., *The number e*, <u>Amer. Math. Monthly</u>, 57 (1950) 591-602; also in T.M. Apostol, <u>Selected Papers on Calculus</u>, M.A.A., 1969, pp. 8-19. [4.1.1]

Dantzig, George B., *Maximization of a linear function of variables subject to linear inequalities*, in T.C. Koopmans, <u>Activity Analysis of Production and Allocation</u>, Wiley, 1951, pp. 339-347. [6.3.1]

Delone, B.N., *Algebra: theory of algebraic equations*, in A.D. Aleksandrov, et al., <u>Mathematics--Its Content, Methods and Meaning</u>, V. 1, M.I.T., 1963, pp. 261-310. [3.6.1]

Dickson, L.E., *The Waring problem and its generalizations*, <u>Bull. Amer. Math. Soc.</u>, 42 (1936) 833-842. [3.3.1]

Drake, Stillman and MacLachlan, James, *Galileo's discovery of the parabolic trajectory*, <u>Scientific American</u>, 232 (March 1975) 102-110, 132. [7.1.0]

Dudley, Underwood, *Who was the first non-Euclidean?*, <u>Math. Spectrum</u>, 6 (1973-74) 41-46. [5.1.0]

Ferguson, Rolfe P., *On Fermat's last theorem*, <u>J. Undergraduate Math.</u>, 6 (1974) 1-14, 85-97; 7 (1975) 35-45. [3.3.2]

Goldstein, L.J., *A history of the prime number theorem*, <u>Amer. Math. Monthly</u>, 80 (1973) 599-615; Correction, 80 (1973) 1115. [3.3.2]

Goldstine, H., <u>The Computer from Pascal to von Neumann</u>, Princeton U. Pr., 1972. [6.5.0]

Grabiner, Judith V., *Is mathematical truth time-dependent?*, <u>Amer. Math. Monthly</u>, 81 (1974) 354-365. [1.5.1]

Haber, Seymour, *Numerical evaluation of multiple integrals*, <u>SIAM Review</u>, 12 (1970) 481-526. [6.4.2]

Harary, Frank, *On the history of the theory of graphs*, in F. Harary, <u>New Directions in the Theory of Graphs</u>, Academic Pr., 1973, pp. 1-17. [3.2.1]

Harary, Frank, *Some historical and intuitive aspects of graph theory*, <u>SIAM Review</u>, 2 (1960) 123-131. [3.2.0]

Heath, F.G., *Origins of the binary code*, <u>Scientific American</u>, 227 (August 1972) 76-83, 124. [6.5.0]

Henkin, Leon, *Are logic and mathematics identical?*, <u>Science</u>, 138 (1962) 788-794. [2.6.0]

Hildebrandt, T.H., *Integration in abstract spaces*, <u>Bull. Amer. Math. Soc.</u>, 59 (1953) 111-139. [4.3.3]

Iyanaga, Shokichi, *Algebraic theory of numbers*, in A. H. Livermore, <u>Science in Japan</u>, AAAS, 1965, pp. 81-113. [3.3.3]

Kendall, David G., *Branching processes since 1873*, <u>J. London Math. Soc.</u>, 41 (1966) 385-406. [6.1.2]

Kendall, David G., *The genealogy of genealogy: branching processes before (and after) 1873*, <u>Bull. London Math. Soc.</u>, 7 (1975) 225-253. [6.1.2]

Lanczos, Cornelius, <u>Space Through the Ages</u>, Academic Pr., 1970. [5.1.1]

McAllister, B.L., *Cyclic elements in topology: a history*, <u>Amer. Math. Monthly</u>, 73 (1966) 337-350. [5.5.2]

Montgomery, David and Quirk, James, *Mathematics in economic theory*, <u>SIAM News</u>, 7:6 (1974) 2-3. [7.4.0]

Morse, Marston, *Trends in analysis*, <u>J. Franklin Inst.</u>, 251 (1951) 33-43. [4.7.3]

Nevanlinna, Rolf, *Methods in the theory of integral and meromorphic functions*, <u>J. London Math. Soc.</u>, 41 (1966) 11-28. [4.4.2]

Ore, Oystein, *Pascal and the invention of probability theory*, <u>Amer. Math. Monthly</u>, 67 (1960) 409-419. [6.1.0]

Polak, E., *An historical survey of computational methods in optimal control*, <u>SIAM Review</u>, 15 (1973) 553-584. [6.3.3]

Randell, Brian, *The Origin of Digital Computers*, Springer-Verlag, 1973. [6.5.0]

Robinson, Abraham, *Some thoughts on the history of mathematics*, Compositio Math., 20 (1968) 188-193. [2.4.1]

Smith, Thomas M., *Some perspectives on the early history of computers*, in Z.W. Pylyshyn, Perspectives on the Computer Revolution, Prentice-Hall, 1970, pp. 7-15. [6.5.0]

Steen, Lynn Arthur, *Highlights in the history of spectral theory*, Amer. Math. Monthly, 80 (1973) 359-381. [4.6.2]

Taylor, Angus E., *Notes on the history of the uses of analyticity in operator theory*, Amer. Math. Monthly, 78 (1971) 331-342. [4.6.3]

Tóth, Imre, *Non-Euclidean geometry before Euclid*, Scientific American, 221 (November 1969) 87-98, 166. [5.1.0]

Turner, James and Kautz, William H., *A survey of progress in graph theory in the Soviet Union*, SIAM Review, 12 (1970) Suppl. pp. 1-68. [3.2.2]

Vandiver, H.S., *Fermat's last theorem--its history and the nature of the known results concerning it*, Amer. Math. Monthly, 53 (1946) 555-578. [3.3.2]

Viscensini, P., *Differential geometry in the nineteenth century*, Scientia, 107 (1972) 661-696. [5.3.3]

Walsh, J.L., *History of the Riemann mapping theorem*, Amer. Math. Monthly, 80 (1973) 270-276. [4.4.3]

Wilder, Raymond L., *History in the mathematics curriculum: its status, quality, and function*, Amer. Math. Monthly, 79 (1972) 479-495. [1.6.1]

Zlot, William Leonard, *The principle of choice in pre-axiomatic set theory*, Scripta Math., 25 (1960) 105-123. [2.1.2]

1.3 BIOGRAPHY

1.3.0 GENERAL

Bell, Eric Temple, Men of Mathematics, Simon and Schuster, 1937.

   This book, consisting of biographies of selected famous mathematicians, is renowned for its author's sprightly and irreverent style.

Grattan-Guinness, Ivor, *A mathematical union: William Henry and Grace Chesholm Young*, Annals of Science, 29 (1972) 105-186.

Grattan-Guinness, Ivor, Joseph Fourier, 1768-1830, M.I.T. Pr., 1972.

Grattan-Guinness, Ivor, *Towards a biography of Georg Cantor*, Annals of Science, 27 (1971) 345-391.

   Based upon a thorough examination of unpublished letters and other documents.

Halmos, Paul R., *"Nicolas Bourbaki"*, Scientific American, 196 (May 1957) 88-99, 174.

Halmos, Paul R., *The legend of John von Neumann*, Amer. Math. Monthly, 80 (1973) 382-394.

Hirolett, J., *Charles Babbage and his computer*, Math. Spectrum, 7 (1974-75) 73-80.

Hoffman, Banesh, Albert Einstein, Creator and Rebel, Viking, 1972.

Infeld, Leopold, Whom the Gods Love, Whittlesey House, 1948.

   Moving biography of Évariste Galois.

Kac, Mark, *Hugo Steinhaus--a reminiscence and a tribute*, Amer. Math. Monthly, 81 (1974) 572-581.

   Kac writes glowingly about the life and career of his teacher and friend.

Kimberling, Clark H., *Emmy Noether*, Amer. Math. Monthly, 79 (1972) 136-149; Addendum, 79 (1972) 755.

   The fascination of this article on the life and work of Emmy Noether is in the skillfully selected extensive quotations from P.S. Alexandroff (never before published), Hermann Weyl, and others. As a by-product of the author's careful preparation of source material he unearthed extensive correspondence of Cantor, Dedekind, Frobenius, and Weber which had been stored in a Philadelphia law office for 33 years. However, he left one stone unturned, as it remained for Dyson to follow up a reference to Einstein and find a 1918 letter by him praising Noether's work; see addendum.

Knuth, Donald E., *George Forsythe and the development of computer science*, Comm. Assoc. Comp. Mach., 15 (1972) 721-726.

Kuratowski, Kazimierz, *Wacław Sierpinski (1882-1969)*, Acta Arith., 21 (1972) 1-5.

   Obituary followed by a summary in English (pp. 7-13) by Schinzel of Sierpinski's papers on number theory and by a bibliography (pp. 15-23).

Lanczos, Cornelius, *William Rowan Hamilton--an appreciation*, Amer. Scientist, 55 (1967) 129-143.

   Survey of Hamilton's work in physics and mathematics.

Lefschetz, Solomon, *Reminiscences of a mathematical immigrant in the United States*, Amer. Math. Monthly, 77 (1970) 344-350.

Meschkowski, Herbert, Ways of Thought of Great Mathematicians, Holden-Day, 1964.

Montgomery, Deane, *Oswald Veblen*, Bull. Amer. Math. Soc., 69 (1963) 26-36.

   Very informative; gives insight into early American mathematics.

Mordell, L.J., *Reflections of a mathematician*, Canad. Math. Congr., 1959.

   Written for the student not interested in the intricacies of mathematics.

Mordell, L.J., *Reminiscences of an octogenarian mathematician*, Amer. Math. Monthly, 78 (1971) 952-961.

   A spritely sequel to *Reflections of a Mathematician*.

Ore, Oystein, Niels Henrik Abel, Mathematician Extraordinary, U. Minn. Pr., 1957; Chelsea, 1974.

Reid, Constance, Hilbert, Springer-Verlag, 1970.

   Impressive biography of David Hilbert (1862-1943), famed mathematician who worked in such varied fields as theory of invariants, theory of algebraic number fields, foundation of geometry, integral equations and physics. The book itself is an interesting account of the life of a prodigy-genius. It traces his education and training quite thoroughly with a minimum of technological jargon and provides interesting philosophical and social insights into both the field of mathematics and the activity of the German intellectual elite in the decades prior to Hitler. Includes many photographs of the Hilbert era and a reprint of the obituary by Hermann Weyl in Bull. Amer. Math. Soc., 50 (1944) 612-654.

Steinhaus, Hugo, *Stefan Banach, 1892-1945*, Scripta Math., 26 (1963) 93-100.

The mathemagician, Time (April 21, 1975) 63.
A biographical sketch of Martin Gardner in honor of his successful April Fool's spoof in Scientific American.

Weyl, Hermann, *Emmy Noether*, Scripta Math., 3 (1935) 201-220.

Wiener, Norbert, Ex-Prodigy, Simon and Schuster, 1953; M.I.T. Pr., 1964.

Wiener, Norbert, I am a Mathematician, Doubleday, 1956; M.I.T. Pr., 1964.

1.3.1 ELEMENTARY

Dieudonné, Jean A., *The work of Nicolas Bourbaki*, Amer. Math. Monthly, 77 (1970) 134-145.
The self-image of Bourbaki is very different from the view that others have of him. This very readable article includes a description of the French mathematical scene circa 1930 and expounds a point of view towards mathematics. Winner of the Lester R. Ford award for expository writing.

Kuhn, Harold W. and Tucker, Albert W., *John von Neumann's work in the theory of games and mathematical economics*, Bull. Amer. Math. Soc., 64 (1958) Suppl. pp. 100-122.
Historical survey with special emphasis on von Neumann.

Mahoney, Michael S., The Mathematical Career of Pierre de Fermat, Princeton, 1973.
A controversial book. See the disparate reviews by André Weil in Bull. Amer. Math. Soc., 79 (1973) 1138-1149 and by D.T. Whiteside in ISIS, 65 (1974) 398-400.

1.3.2 ADVANCED

Tate, J., *The 1974 Fields Medals (I): An algebraic geometer*, Science, 186 (1974) 39-40.

David Mumford and his contributions.

Ulam, Stanislaw M., *John von Neumann, 1903-1957*, Bull. Amer. Math. Soc., 64 (1958) Suppl. pp. 1-49.

Survey of his work, of varying complexity.

1.3.3 RESEARCH

Almgren, F.J., Jr. and Montgomery, H., *The 1974 Fields Medals (II): An analyst and number theorist*, Science, 186 (1974) 130-131.

Description of the recent work of Enrico Bombieri.

RELATED REFERENCES

Bell, Eric Temple, *Gauss and the early development of algebraic numbers*, National Math. Magazine, 18 (1944) 188-204, 219-233. [1.2.2]

Bell, Eric Temple, *Newton after three centuries*, Amer. Math. Monthly, 49 (1942) 553-575. [1.2.1]

Bochner, Salomon, *Mathematical reflections*, Amer. Math. Monthly, 81 (1974) 827-852. [1.2.1]

Brown, Brenda W. and Brown, Robert F., *Hardy's "Apology": classic essay or cloistral clowning?*, Delta, 3:3 (Spring 1973) 1-10. [1.5.0]

Courant, Richard, *Gauss and the present situation of the exact sciences*, in T.L. Saaty and F.J. Weyl, The Spirit and the Uses of the Mathematical Sciences, McGraw-Hill, 1969, pp. 141-155. [1.5.0]

Hardy, G.H., A Mathematician's Apology, Cambridge Univ. Pr., 1940, 1967; excerpted in J.R. Newman, The World of Mathematics, V. 4, Simon and Schuster, 1956, pp. 2027-2038. [1.5.0]

Lebesgue, Henri, *The development of the integral concept*, in H. Lebesgue, Measure and the Integral, Holden-Day, 1966, pp. 178-194. [4.3.2]

1.4 Role of Mathematics

1.4.0 GENERAL

Birkhoff, George D., *The mathematical nature of physical theories*, Amer. Scientist, 31 (1943) 281-310.

Bronowski, Jacob, *The logic of the mind*, Amer. Scientist, 54 (1966) 1-14.

> Holds that the logical theorems of Gödel "reach decisively into the systemization of empirical science", and that "it is the language that we use in describing nature that imposes...both the form and limitations of the laws that we find." Bronowski then goes on to compare the situations of science and literature, concluding that they differ in the extent to which self-reference enters their languages. Integrative and interdisciplinary.

Bronowski, Jacob, *The music of the spheres*, in J. Bronowski, The Ascent of Man, Little, Brown, 1973, pp. 154-187.

> Wide-ranging essay by a prominent humanist and bridger of the two-cultures chasm--himself a mathematician. Discusses Pythagoras and the nature of logical proof, contributions of the Arabs (the Astrolabe, Hindu-Arabic numerals, geometrical patterns at the Alhambra), structure of crystals, the significance of the development of perspective in painting, too brief mention of the calculus. Not deep, but broad and a little profound. Lavishly illustrated; the essay is meant to accompany Bronowski's related TV film.

Browder, Felix E., *Is mathematics relevant? And if so, to what?*, Univ. of Chicago Magazine, 67:3 (Spring 1975) 11-16; also appears as *The relevance of mathematics*, Amer. Math. Monthly, 83 (1976) 249-254.

De Broglie, Louis, *The role of mathematics in the development of contemporary theoretical physics*, in F. LeLionnais, Great Currents of Mathematical Thought, V. 2, Dover, 1971, pp. 78-93.

Fehr, Howard F., *Value and the study of mathematics*, Scripta Math., 21 (1955) 49-53.

> Intrinsic and extrinsic values.

Lieber, Lillian R., Human Values and Science, Art and Mathematics, Norton, 1961.

Morse, Marston, *Mathematics in our culture*, in T.L. Saaty and F.J. Weyl, <u>The Spirit and Uses of the Mathematical Sciences</u>, McGraw-Hill, 1969, pp. 105-120.

> Many examples of mathematics in natural sciences, social sciences, art, music and poetry.

Stone, Marshall H., *Mathematics and the future of science*, <u>Bull. Amer. Math. Soc.</u>, 63 (1957) 61-76.

> Excellent exposition of the role of pure mathematics in applied science, with examples of how mathematics is applied.

Whitehead, Alfred North, *Mathematics and liberal education*, in A.N. Whitehead, <u>Essays in Science and Philosophy</u>, Philosophical Library, 1947, pp. 175-188.

Wigner, Eugene P., *The unreasonable effectiveness of mathematics in the natural sciences*, <u>Comm. Pure Appl. Math.</u>, 13 (1960); also in T.L. Saaty and F.J. Weyl, <u>The Spirit and Uses of the Mathematical Sciences</u>, McGraw-Hill, 1969, pp. 123-140; in <u>Studies in Mathematics</u>, V. 16, SMSG, 1967, pp. 31-44; and in E.P. Wigner, <u>Symmetries and Reflections: Scientific Essays of Eugene P. Wigner</u>, Indiana U. Pr., 1967, pp. 222-237.

> Interesting examples of uncanny relations between mathematics and physics..."a wonderful gift which we neither understand nor deserve."

Wilder, Raymond L., *Trends and social implications of research*, <u>Bull. Amer. Math. Soc.</u>, 75 (1969) 891-906.

1.4.1 ELEMENTARY

Cohen, Hirsh, *Mathematical applications, computation, and complexity*, <u>Quarterly of Applied Math.</u>, 30 (1972) 109-121.

> Cohen discusses the future of applied mathematics, and suggests some fields outside the physical sciences in which he hopes to see mathematics applied. In particular, he outlines various types of problems in computer science which he thinks applied mathematicians can profitably investigate, including algorithmic complexity, operating system modeling, and data structuring.

Courant, Richard, *Mathematics in the modern world*, <u>Scientific American</u>, 211 (September 1964) 40-49, 269; also in M. Kline, <u>Mathematics in the Modern World</u>, Freeman, 1968, pp. 19-27, 394.

> Numerous examples of pure and applied mathematics, accenting their inseparability.

Greenspan, H.P., *Applied mathematics as a science*, Amer. Math. Monthly, 68 (1961) 872-880.

   Plea for more emphasis on applied mathematics. Includes several interesting examples of calculus in action.

Schwartz, Jacob T., *The pernicious influence of mathematics on science*, in E. Nagel, P. Suppes and A. Tarski, Logic, Methodology and Philosophy of Science, Stanford, 1962, pp. 356-360.

Weyl, Hermann, Symmetry, Princeton U. Pr., 1952; excerpted in J.R. Newman, The World of Mathematics, V. 1, Simon and Schuster, 1956, pp. 671-724.

   A gentle yet profound exposition of the major ideas that led to the theory of groups, demonstrating the work of the mathematical intellect in the evolution of intuitive concepts into grand systems of abstract ideas.

RELATED REFERENCES

Barbut, Marc, *Does the majority ever rule?*, Portfolio and Art News Annual, 4 (1961) 79-83, 161-168. [7.3.0]

Dyson, Freeman J., *Mathematics in the physical sciences*, Scientific American, 211 (September 1964) 128-146, 269-270; also in M. Kline, Mathematics in the Modern World, Freeman, 1968, pp. 249-257, 401; and in The Mathematical Sciences--A Collection of Essays for COSRIMS, M.I.T., 1969, pp. 97-115. [7.1.0]

Dyson, Freeman J., *Missed opportunities*, Bull. Amer. Math. Soc., 78 (1972) 635-652. [7.1.2]

Hamming, Richard W., *Impact of computers*, Amer. Math. Monthly, 72 (1965) Suppl. pp. 1-7. [6.5.0]

Hamming, Richard W., *One man's view of computer science*, J. Assoc. Comp. Mach., 16 (1969) 3-12. [6.5.0]

Henrici, Peter, *Reflections of a teacher of applied mathematics*, Quarterly of Applied Math., 30 (1972) 31-39. [1.6.1]

Klamkin, Murray S., *On the ideal role of an industrial mathematician and its educational implications*, Educ. Studies Math., 3 (1970-71) [1.6.0]

244-269; also in Amer. Math. Monthly, 78
(1971) 53-76.

Klamkin, Murray S., *The teaching of mathematics* [1.6.1]
so as to be useful, Educ. Studies Math., 1
(1968-69) 126-160.

Lighthill, M.J., *The art of teaching the art* [1.6.0]
of applying mathematics, Math. Gazette, 55
(1971) 249-270.

Pollak, Henry O., *How can we teach applications* [1.6.1]
of math?, Educ. Studies Math., 2 (1969-70) 393-
404.

Pollak, Henry O., *On some of the problems of* [1.6.1]
teaching applications of mathematics, Educ.
Studies Math., 1 (1968-69) 24-30.

Shubnikov, A.V. and Koptsik, V.A., Symmetry [3.5.1]
in Science and Art, Plenum Pr., 1974.

Stevens, Peter S., Patterns in Nature, [5.1.0]
Atlantic-Little, Brown, 1974.

1.5 NATURE OF MATHEMATICS

1.5.0 GENERAL

Adler, Alfred, *Reflections--mathematics and creativity*, New Yorker, 47 (February 19, 1972) 39-45.

> Mathematics as a search for analogies. Contains many controversial generalizations. See commentary by K.O. May, Notices Amer. Math. Soc., 19:4 (June 1972) 207.

Bourbaki, Nicolas, *The architecture of mathematics*, Amer. Math. Monthly, 57 (1950) 221-232, in F. LeLionnais, Great Currents of Mathematical Thought, V. 1, Dover, 1971, pp. 23-36.

> Morphology of mathematics--very general. Is there just one mathematic, or many mathematics? Rather heavy reading, but technically simple.

Brown, Brenda W. and Brown, Robert F., *Hardy's "Apology": classic essay or cloistral clowning?*, Delta, 3:3 (Spring 1973) 1-10.

Cartwright, Mary L., *Mathematics and thinking mathematically*, Amer. Math. Monthly, 77 (1970) 20-28.

> On the distinction between pure and applied mathematics. Discusses, with numerous examples, whether a work is any more mathematical because it is abstract and studied strictly for its own sake.

Cartwright, Mary L., *The mathematical mind*, Math. Spectrum, 2 (1969-70) 37-45.

> Illuminates for a general audience two questions which are asked by every mathematician: "Would I achieve more if I changed my methods of working?" and "What constitutes good mathematics and how does one recognize it?"

Courant, Richard, *Gauss and the present situation of the exact sciences*, in T.L. Saaty and F.J. Weyl, The Spirit and the Uses of the Mathematical Sciences, McGraw-Hill, 1969, pp. 141-155.

Dieudonné, Jean A., *Mathematics*, Collier's Encyclopedia, 1976, V. 15, pp. 541-552.

Eilenberg, Samuel, *The algebraization of mathematics*, in The Mathematical Sciences--A Collection of Essays for COSRIMS, M.I.T., 1969, pp. 153-160.

> The title is more formidable than the article, since the author has elected to be very gentle in his mathematical demands on the

reader. He mentions some significant parts of algebra and indicates briefly the role of modern algebra in other areas.

Ficken, F.A., *Mathematics and the layman*, Amer. Scientist, 52 (1964) 419-430.
> Tries to convey the flavor and special appeal of mathematics by exploring the nature of mathematical activity.

Hadamard, Jacques, The Psychology of Invention in the Mathematical Field, Princeton U. Pr., 1949.
> A brief classic, inspired by Poincaré's lectures. Hadamard focuses on four stages of creative mathematical thought--intense concentration, relaxation or distraction, sudden insight, careful reconsideration.

Hahn, Hans, *The crisis in intuition*, in J.R. Newman, The World of Mathematics, V. 3, Simon and Schuster, 1956, pp. 1956-1976.
> Attacks Kant's claim that mathematics relies on intuition by exhibiting counterintuitive examples (especially from calculus).

Halmos, Paul R., *Mathematics as a creative art*, Amer. Scientist, 56 (1968) 375-389.
> An excellent article which explains what mathematicians do, distinguishes between pure and applied mathematics ("mathology" vs. "mathophysics"), and draws upon analogies with the other arts to convince the reader that mathematics is a creative art.

Hardy, G.H., A Mathematician's Apology, Cambridge U. Pr., 1940; 1967; excerpted in J.R. Newman, The World of Mathematics, V. 4, Simon and Schuster, 1956, pp. 2027-2038.
> Brief, stunning, controversial and famous defense of pure mathematics by England's greatest number theorist. The work conveys an attitude of haughty elitism ("most people can do nothing well") mixed with chauvinism ("Archimedes will be remembered when Aeschylus is forgotten"). Hardy bases his justification for mathematics on its harmlessness, the permanence of its achievement, and the pleasure that doing it gives. The 1967 edition has a lengthy foreword by C.P. Snow that is of equal interest.

Helitzer, Florence, *A conversation with three mathematicians*, University: A Princeton Quarterly, 59 (Winter 1974) 1-5, 28-30.
> Engaging interview with William Browder, Donald Spencer and Michael Reed by the assistant director of the Princeton News Bureau.

Hilton, Peter J., *The art of mathematics*, Univ. of Birmingham, 1960.

  An inaugural lecture that moves eloquently from a general analysis of the explosive growth of mathematical research to an exposition of the work of the 1958 Fields medalists--K.F. Roth and R. Thom.

Iliev, L., *Mathematics as the science of models*, Russian Math. Surveys, 27:2 (1972) 181-189.

  Regards mathematics as the construction of models of real phenomena. Argues that a major push for the creation of new mathematical structures comes today from such fields as computer technology and cybernetics; the possibility of obtaining new mathematical structures, not isomorphic to existing ones, is an exciting prospect for mathematics.

Jones, Landon Y., Jr., *Mathematicians: They're special*, Think, 40:4 (1974) 32-35.

Lefschetz, Solomon, *The structure of mathematics*, Amer. Scientist, 38 (1950) 105-111.

  Non-technical exposition of algebra and topology as the two "polar caps" of the "distant planet" of mathematics.

Murray, Francis J. and Ford, Lester R., *Mathematics as a calculatory science*, Encyclopaedia Britannica, 15th ed., 1974, Macropaedia V. 11, pp. 671-696.

  "Mathematics as a calculation science is then concerned with the ingredients of mathematics that characterize the automated form of the reasoning process and the methods of thought that rely upon such things as physical models..." Topics such as numerical notations, mathematical models, graphical procedures, and analog and digital computers are included.

Newman, M.H.A., *What is mathematics? New answers to an old question*, Math. Gazette, 43 (1959) 161-171.

  Easy, brief discussion of the meaning of mathematics.

Poincaré, Henri, *Mathematical creation*, Scientific American, 179 (August 1948) 54-57; also in M. Kline, Mathematics in the Modern World, Freeman, 1968, pp. 14-17; and in J.R. Newman, The World of Mathematics, V. 4, Simon and Schuster, 1956, pp. 2041-2050.

  Mathematical creation depends on subconscious intuition of mathematical order, thus it depends heavily on "sudden inspiration." According to Poincaré, this explains how it is possible for many otherwise rational people to fail to understand mathematics even though it is completely logical.

Rényi, Alfréd, *A Socratic dialogue on mathematics*, Canad. Math. Bull., 7 (1964) 441-462; also in A. Rényi, Dialogues on Mathematics, Holden-Day, 1967, pp. 3-25.

   The object and utility of mathematics. Its nature and its concepts investigated in a distinctive and delightful style.

Salmon, Wesley C., *Confirmation*, Scientific American, 228 (May 1973) 75-83, 120.

   The logic of the process of testing, confirming, and disconfirming hypotheses--with probabilities playing a major role.

Spohn, William G., Jr., *Can mathematics be saved?*, Notices Amer. Math. Soc., 16 (1969) 890-894.

Stein, Sherman K., *The mathematician as an explorer*, Scientific American, 204 (May 1961) 148-158, 206.

   Fascinating elementary account of how a mathematical investigation of an ancient Indian drummer's memory word leads to the travelling salesman's problem.

Stone, Marshall H., *The future of mathematics*, J. Math. Soc. Jap., 9 (1957) 493-507.

Suppes, P., *A comparison of the meaning and uses of models in mathematics and the empirical sciences*, Synthese, 12 (1960) 287-301.

Szabó, Árpád, *The transformation of mathematics into a deductive science and the beginnings of its foundations on definitions and axioms*, Scripta Math., 27 (1964) 28-48A, 113-139.

   A new approach to answering the question of how mathematics first became a deductive science.

von Neumann, John, *The mathematician*, in R.B. Heywood, The Works of the Mind, U. of Chicago Pr., 1947, pp. 180-196; also in J.R. Newman, The World of Mathematics, V. 4, Simon and Schuster, 1956, pp. 2053-2063.

   John von Neumann, one of the foremost mathematicians of our time, made fundamental contributions to several branches of mathematics. In this essay he suggests that the criteria of mathematical success are "almost entirely aesthetical," but mathematical inspiration comes from natural experience.

Weidman, Donald R., *Emotional perils of mathematics*, Science, 149 (1965) 1048.

   An extensive letter commenting on four emotional problems unique to mathematicians.

Weissinger, Johannes, *The characteristic features of mathematical thought*, in T.L. Saaty and F.J. Weyl, <u>The Spirit and Uses of the Mathematical Sciences</u>, McGraw-Hill, 1969, pp. 9-27.

> Well written and profusely illustrated discussion of mathematics as the science of structure.

Weyl, Hermann, *Insight and reflection*, <u>Studia Philosophica</u>, 15 (1955); also in T.L. Saaty and F.J. Weyl, <u>The Spirit and Uses of the Mathematical Sciences</u>, McGraw-Hill, 1969, pp. 281-301.

Weyl, Hermann, *The mathematical way of thinking*, <u>Science</u>, 92 (1940) 437-446; also in <u>Studies in the History of Science</u>, U. Penn. Pr., 1941, pp. 103-123.

Whitehead, Alfred North, *Mathematics as an element in the history of thought*, in J.R. Newman, <u>The World of Mathematics</u>, V. 1, Simon and Schuster, 1956, pp. 402-416.

> The first three sentences of this provocative article: "[T]he science of pure mathematics, in its modern developments, may claim to be the most original creation of the human spirit. Another claimant for this position is music. But we will put aside all rivals, and consider the ground on which such a claim can be made for mathematics."

Wilder, Raymond L., *The role of the axiomatic method*, <u>Amer. Math. Monthly</u>, 74 (1967) 115-127; also in <u>Math. Teaching</u>, 41 (1967) 32-40.

> Expository, historical. Mostly about logic and set theory. Long bibliography.

1.5.1 ELEMENTARY

Grabiner, Judith V., *Is mathematical truth time-dependent?*, <u>Amer. Math. Monthly</u>, 81 (1974) 354-365.

> The author examines the changes in attitude toward mathematical rigor between the 18th and 19th centuries, and the possible reasons for these changes. The discussion includes examples of how 18th-century methods--e.g., in approximation theory--became the technical basis for 19th-century rigor.

Lehmer, D.H., *Mechanized mathematics*, <u>Bull. Amer. Math. Soc.</u>, 72 (1966) 739-750.

> Influence of mechanical aids on mathematics. Unusual examples.

Taylor, Angus E., *Some aspects of mathematical research*, Amer. Scientist, 35 (1947) 211-223.

> How do you go about carrying on mathematical research? An attempt to answer this perennial question for non-mathematicians.

Wilder, Raymond L., *The nature of mathematical proof*, Amer. Math. Monthly, 51 (1944) 309-323.

> Surveys methods of proof. Argues that proof is only a "testing process" which we apply to suggestions of intuition.

Wilder, Raymond L., *The origin and growth of mathematical concepts*, Bull. Amer. Math. Soc., 59 (1963) 423-448.

> Aims at showing that simultaneous discoveries are not coincidences but necessities. Heaviest detail is on curves and topology.

1.5.2 ADVANCED

Köthe, G. and Ballier, F., *The changing structure of modern mathematics*, in H. Behnke, et al., Fundamentals of Mathematics, V. 3, M.I.T., 1974, pp. 505-528.

Lakatos, Imre, *Proofs and refutations*, Brit. J. Phil. Science, 14 (1963-64) 1-25, 120-139, 221-245, 296-342.

> Provocative exposition in dialog form of the dynamics by which mathematical ideas develop through the interplay of proofs and counterexamples. Synthesis of ideas of George Pólya in heuristics and Karl Popper in philosophy within the framework of the history of the Euler characteristic for polyhedra.

Nalimov, V.V., *Logical foundations of applied mathematics*, Synthese, 27 (1974) 211-250.

1.5.3 RESEARCH

Kac, Mark, *On applying mathematics: reflections and examples*, Quarterly of Applied Math., 30 (1972) 17-29.

> In the context of a lecture on the nature of applied mathematics, the author gives brief discussions of the Ising model and the spherical model in ferromagnetism. He also describes how consideration of a Bose-Einstein gas in a "rotating bucket" provided new perspective on the distribution of the eigenvalues of the Laplacian.

RELATED REFERENCES

Adams, Ernest W., *On the nature and purpose of measurement*, Synthese, 16 (1966) 125-169; also in B. Lieberman, Contemporary Problems in Statistics, Oxford U. Pr., 1971, pp. 74-92. [6.2.0]

Hahn, Hans, *Geometry and intuition*, Scientific American, 190 (April 1954) 84-91, 108; also in M. Kline, Mathematics in the Modern World, Freeman, 1968, pp. 184-188, 399. [5.1.0]

Hamming, Richard W., *Intellectual implications of the computer revolution*, Amer. Math. Monthly, 70 (1963) 4-11; also in T.L. Saaty and F.J. Weyl, The Spirit and Uses of the Mathematical Sciences, McGraw-Hill, 1969, pp. 188-199; in Studies in Mathematics, V. 16, SMSG, 1967, pp. 45-52; and in Z.W. Pylyshyn, Perspectives on the Computer Revolution, Prentice-Hall, 1970, pp. 370-377. [6.5.0]

Kemeny, John G., *Teaching the new mathematics*, Atlantic Monthly, October 1962; also in J.G. Kemeny, Random Essays on Mathematics, Education and Computers, Prentice-Hall, 1964, pp. 27-34. [1.6.0]

Klamkin, Murray S. and Newman, D.J., *The philosophy and applications of transform theory*, SIAM Review, 3 (1961) 10-36. [7.1.2]

Knuth, Donald E., *Computer programming as an art*, Comm. Assoc. Comp. Mach., 17 (1974) 667-673. [6.5.0]

Richards, Ian, *Impossibility*, Math. Magazine, 48 (1975) 249-262. [3.6.1]

1.6 PEDAGOGY

1.6.0 GENERAL

Brainerd, Charles J., *The origins of number concepts*, <u>Scientific American</u>, 228 (March 1973) 100-109, 128.

"Experiments with children indicate that they first become aware of numbers in terms of ordered sequences and only later in terms of quantities."

Dieudonné, Jean A., *Should we teach "modern" mathematics?*, <u>Amer. Scientist</u>, 61 (1973) 16-19.

Prompted by René Thom's *"Modern" mathematics: an educational and philosophic error?* (<u>Amer. Scientist</u>, 59 (1971) 695-699), Dieudonné sets forth his own opinions under three headings: rigor and axiomatization, mathematics at the university level, and mathematics at the school level.

Halmos, Paul R., *How to talk mathematics*, <u>Notices Amer. Math. Soc.</u>, 21 (1974) 155-158.

Halmos, Paul R., *How to write mathematics*, <u>L'Enseignement Math.</u>, 16 (1970) 123-152.

This essay, addressed to "mathematics students who are near the beginning of their thesis work" and "colleagues whose ways can stand mending", is a manual of style which is devoted to the peculiar problems of writing mathematics. The topics range from overall organization to minute but helpful hints on punctuation. The essay itself is a sample of the author's flair for good writing.

Hilton, Peter J., *The survival of education*, <u>Educ. Tech.</u>, 13:11 (November 1973) 12-16.

Jones, Phillip S., *The history of mathematical education*, <u>Amer. Math. Monthly</u>, 74 (1967) Suppl. pp. 38-55.

Interesting survey of trends in mathematical education beginning with the ancients.

Kemeny, John G., *Teaching the new mathematics*, <u>Atlantic Monthly</u>, October 1962; also in J.G. Kemeny, <u>Random Essays on Mathematics, Education and Computers</u>, Prentice-Hall, 1964, pp. 27-34.

What is modern mathematics? The search for pattern, not a particular subject.

Klamkin, Murray S., *On the ideal role of an industrial mathematician and its educational implications*, Educ. Studies Math., 3 (1970-71) 244-269; also in Amer. Math. Monthly, 78 (1971) 53-76.

> Narrow training of Ph.D.'s, stages in problem solution, remarks on geometry.

Lazarus, Mitchell, *Mathophobia: some personal speculations*, Nat. Elem. Principal, 53:2 (Jan.-Feb. 1974) 16-22.

Lighthill, M.J., *The art of teaching the art of applying mathematics*, Math. Gazette, 55 (1971) 249-270.

> "The key problems are of a communications and linguistic nature."

Meserve, Bruce E., *New mathematics*, Encyclopedia Americana, 1976, V. 20, pp. 202-205.

Nevanlinna, Rolf, *Reform in teaching mathematics*, Amer. Math. Monthly, 73 (1966) 451-464.

> A noted mathematician's views on "certain characteristic features in the modern development of mathematical research" as background for his discussion of teaching. He stresses the impact on modern mathematics of ideas from Euclidean geometry.

Ordman, Edward T., *One and one is nothing: liberating mathematics*, Soundings, 56 (1973) 164-181.

> "The purpose of this essay is to give some rather personal impressions of what mathematics is, of why mathematics is a liberal art, and of what sorts of things might go to make up a 'liberating' mathematics course... Many mathematicians feel...that our nonmathematical colleagues have little idea of what mathematics is...or what mathematics is doing in the liberal arts. I hope that by actually discussing some specific pieces of mathematics [a theorem on mazes, Euler's formula, Gödel incompleteness]...I can cast some limited amount of light on these issues." The author's claim is that mathematics is a liberal art because it reveals something about the human condition.

Piaget, Jean, Genetic Epistemology, Columbia U. Pr., 1970.

> Short and readable. He finds that in children the notion of counting develops simultaneously with the notion of 1-1 map; the notion of speed is *prior* to the notion of time (duration).

Pólya, George, How to Solve It, Princeton U. Pr., 1945; excerpted in J.R. Newman, The World of Mathematics, V. 3, Simon and Schuster, 1956, pp. 1980-1992.

> Includes an excellent exposition of mathematical induction, setting up equations and working backwards. Pólya describes his book as heuristic. Very readable!

Thom, René F., *"Modern" mathematics: an educational and philosophic error?*, <u>Amer. Scientist</u>, 59 (1971) 695-699.

> The author contrasts algebra and set theory (in the sense of the "new math") with geometry as a vehicle for presenting mathematics in the schools, and comes out forcefully in favor of geometry. He also offers various provocative opinions on the nature of mathematical thought.

1.6.1 ELEMENTARY

Engel, Arthur, *The relevance of modern fields of applied mathematics for mathematical education*, <u>Educ. Studies Math.</u>, 2 (1969-70) 257-269.

> A short prose tour through operations research, mathematical bioscience, algorithms. Emphasizes projects to engage students.

Freudenthal, Hans, <u>Mathematics as an Educational Task</u>, D. Reidel, 1973.

Henrici, Peter, *Reflections of a teacher of applied mathematics*, <u>Quarterly of Applied Math.</u>, 30 (1972) 31-39.

Klamkin, Murray S., *The teaching of mathematics so as to be useful*, <u>Educ. Studies Math.</u>, 1 (1968-69) 126-160.

> Emphasizes horizontal learning, wealth of examples, outline of horizontal problem-solving course. Excellent.

Kline, Morris, *Logic versus pedagogy*, <u>Amer. Math. Monthly</u>, 77 (1970) 264-282.

Pollak, Henry O., *How can we teach applications of math?*, <u>Educ. Studies Math.</u>, 2 (1969-70) 393-404.

> Debunks phony applications. Lots of examples.

Pollak, Henry O., *On some of the problems of teaching applications of mathematics*, <u>Educ. Studies Math.</u>, 1 (1968-69) 24-30.

> Isolates five false folk theorems, misconceptions about applied mathematics. Contains seven compelling down-to-earth applied problems.

Pólya, George, <u>Mathematical Discovery</u>, Wiley, V. 1, 1962, V. 2, 1965.

Pólya, George, <u>Mathematics and Plausible Reasoning</u>: V. I, <u>Induction and Analogy in Mathematics</u>; V. II, <u>Patterns of Plausible Inference</u>, Princeton U. Pr., 1954.

Wilder, Raymond L., *History in the mathematics curriculum: its status, quality, and function*, <u>Amer. Math. Monthly</u>, 79 (1972) 479-495.

Winner of the Lester R. Ford award for expository writing.

RELATED REFERENCES

Hilton, Peter J., *Topology in the high school*, [5.7.2]
<u>Educ. Studies Math.</u>, 3 (1970-71) 436-453.

Minsky, Marvin L., *Form and content in computer science*, <u>J. Assoc. Comp. Mach.</u>, 17 (1970) [6.5.0]
197-215.

Wilder, Raymond L., *The nature of mathematical proof*, <u>Amer. Math. Monthly</u>, 51 (1944) 309-323. [1.5.1]

1.7 MISCELLANY

1.7.0 GENERAL

Abbott, Edwin A., Flatland--A Romance of Many Dimensions, Little, Brown, 1928; Dover, 1952.
   A unique and well-known classic, originally published in 1884.

Ball, W.W. Rouse and Coxeter, H.S. MacDonald, Mathematical Recreations & Essays, Twelfth Edition, U. of Toronto Pr., 1974.

Birkhoff, George D., Aesthetic Measure, Harvard U. Pr., 1933.
   Aesthetic theory, polygonal forms, ornaments and tiling.

De Morgan, Augustus, A Budget of Paradoxes, Open Court, 1872; 1915; excerpted in J.R. Newman, The World of Mathematics, V. 4, Simon and Schuster, 1956, pp. 2369-2382.

Gardner, Martin, Martin Gardner's Sixth Book of Mathematical Games from Scientific American, Freeman, 1971.

Gardner, Martin, Mathematical Carnival, Knopf, 1975.

Gardner, Martin, Mathematics, Magic and Mystery, Dover, 1956.

Gardner, Martin, New Mathematical Diversions from Scientific American, Simon and Schuster, 1966.

Gardner, Martin, The Numerology of Dr. Matrix, Simon and Schuster, 1967.

Gardner, Martin, The Scientific American Book of Mathematical Puzzles and Diversions, Simon and Schuster, 1959.

Gardner, Martin, The Second Scientific American Book of Mathematical Puzzles and Diversions, Simon and Schuster, 1961.

Gardner, Martin, The Unexpected Hanging, and Other Mathematical Diversions, Simon and Schuster, 1969.

Huntley, H.E., The Divine Proportion: A Study in Mathematical Beauty, Dover, 1970.

Jones, Landon Y., Jr., *Bad days on Mount Olympus*, Atlantic

Monthly, 233 (February 1974) 37-46, 51-53.
Recent controversies at the Institute for Advanced Study.

Lieber, Lillian R., Mits, Wits, and Logic, Norton, 1947; 1954; 1960.

Lieber, Lillian R., Take a Number, Ronald Pr., 1946.

Lieber, Lillian R., The Education of T.C. Mits, Norton, 1942; 1944.

1.7.1 ELEMENTARY

Kürschák, József, Hungarian Problem Book I, New Math. Libr., No. 11, Random House, 1963; M.A.A., 1975.

Kürschák, József, Hungarian Problem Book II, New Math. Libr., No. 12, Random House, 1963; M.A.A., 1975.

Maxwell, E.A., Fallacies in Mathematics, Cambridge U. Pr., 1963.

Salkind, Charles T., The Contest Problem Book, New Math. Libr., No. 5, Random House, 1961; M.A.A., 1975.

Salkind, Charles T., The M.A.A. Problem Book II, New Math. Libr., No. 17, Random House, 1966; M.A.A., 1975.

Salkind, Charles T. and Earl, James M., The M.A.A. Problem Book III, New Math. Libr., No. 25, Random House, 1973; M.A.A., 1975.

1.7.2 ADVANCED

Dudley, Patricia L., et al., *Further techniques in the theory of big game hunting*, Amer. Math. Monthly, 75 (1968) 896-897.

Morphy, Otto, *Some modern mathematical methods in the theory of lion hunting*, Amer. Math. Monthly, 75 (1968) 185-187.

Pétard, H., *A contribution to the mathematical theory of big game hunting*, Amer. Math. Monthly, 45 (1938) 446-447.
Sophisticated humor.

2.1 SET THEORY

2.1.0 GENERAL

Hahn, Hans, *Infinity*, in J.R. Newman, The World of Mathematics, V. 3, Simon and Schuster, 1956, pp. 1593-1611.

Hahn, Hans, *Is there an infinity?*, Scientific American, 187 (November 1952) 76-84, 104.

Kasner, Edward and Newman, James R., *Paradox lost and paradox regained*, in J.R. Newman, The World of Mathematics, V. 3, Simon and Schuster, 1956, pp. 1936-1955.
  A representative sample of paradoxes in geometry, set theory, arithmetic and logic. Provides excellent perspective on the role of paradoxes in mathematics.

Lieber, Lillian R., Infinity, Holt, Rinehart and Winston, 1953; 1964.

Smullyan, Raymond M., *The continuum hypothesis*, in The Mathematical Sciences--A Collection of Essays for COSRIMS, M.I.T., 1969, pp. 252-260.
  Description of the current status of the continuum hypothesis and the axiom of constructibility in clear but elementary terms.

Vilenkin, N. Ya, Stories about Sets, Academic Pr., 1968.
  An exposition of "corridor" mathematics--in this case, mathematics learned in the corridors of Moscow State University by the author while a beginning student--sugar coated for this popularization. Cardinality is developed far enough to discuss the continuum hypothesis and various unusual counterintuitive examples.

Zippin, Leo, Uses of Infinity, New Math. Libr., No. 7, Random House, 1962; M.A.A., 1975.
  "The philosopher and theologian are conscious of infinity, but from the mathematician's point of view they do not use it so much as admire it." The author proceeds to explain what he means by "using" infinity, ranging from ancient geometrical constructions to the Bolzano-Weierstrass principle.

2.1.1 ELEMENTARY

Arnold, B.H., *Set theory*, Encyclopedia Americana, 1976, V. 24, pp. 588-591.

Black, Max, *The elusiveness of sets*, Review of Metaphysics, 24 (1971) 614-636.

Cohen, Paul J. and Hersh, Reuben, *Non-Cantorian set theory*, Scientific American, 217 (December 1967) 104-116, 160; also in M. Kline, Mathematics in the Modern World, Freeman, 1968, pp. 212-220, 400.

> Anyone attempting to explain the subtleties of the roles of the axiom of choice and the continuum hypothesis in mathematical logic will naturally think of using the analogy with Euclidean and non-Euclidean geometry. Cohen and Hersh draw this analogy in a particularly illuminating way to explain in a simplified fashion how the continuum hypothesis may fail.

Hashisaki, Joseph and Stoll, Robert R., *Set theory*, Encyclopaedia Britannica, 15th ed., 1974, Macropaedia V. 16, pp. 569-575.

> A survey of axiomatic set theory, with detailed discussion of the Zermelo-Fraenkel axioms and the von Neumann-Bernays-Gödel axioms.

Hilbert, David, *On the infinite*, in P. Benacerraf and H. Putnam, Philosophy of Mathematics, Prentice-Hall, 1964, pp. 134-151.

> Hilbert's opinions and hopes for establishing mathematics on absolutely secure foundations, written six years before Gödel's proof destroyed all such hopes forever.

Kolata, Gina Bari, *Foundations of mathematics: ties to infinite games*, Science, 188 (1975) 923-924.

> Popular account of background and implications of Martin's result that Borel sets are determined.

Smullyan, Raymond M., *The continuum problem*, The Encyclopedia of Philosophy, Macmillan, 1967, V. 2, pp. 207-212.

2.1.2 ADVANCED

Abian, A., *On inaccessible cardinal numbers*, Arch. Math. Logik, 12 (1969) 99-103.

> A little textbook on different kinds of inaccessible cardinals.

Cohen, Paul J., *Comments on the foundations of set theory*, in D. Scott, Axiomatic Set Theory, Part 1, Amer. Math. Soc., 1971, pp. 9-15.

Gödel, Kurt, *What is Cantor's continuum problem?*, Amer. Math. Monthly, 54 (1947) 515-525; also in P. Benacerraf and H. Putnam, Philosophy of Mathematics, Prentice-Hall, 1964, pp. 258-273.

>Extensive survey presuming thorough grounding in the theory of cardinal numbers. Now obviously dated, yet still of considerable value. A rare and fascinating glimpse at the opinions of the creator of modern logic.

Monk, J. Donald, *On the foundations of set theory*, Amer. Math. Monthly, 77 (1970) 703-711.

>Brief survey of some recent accomplishments in set theory (e.g., theorems of Gödel and Cohen, results of Solovay pertaining to questions of the existence of Lebesgue non-measurable sets) together with their philosophical and pedagogical impact.

Rubin, Jean E., *Finite sets*, Math. Magazine, 46 (1973) 183-192.

>"Since the notion of a finite set is intuitively so simple, it is aesthetically desirable to define it and derive its properties without using such a powerful tool as the axiom of infinity. Why should an infinite set be required for the study of finite sets? ...[Tarski] developed [in 1924] a theory of finite sets which did not depend on the axiom of infinity." The author follows Tarski to show the equivalence of seven notions of finiteness, and also the equivalence of three notions of Dedekind finiteness (S is not equipollent to any proper subset of itself). The equivalence of finiteness and Dedekind finiteness requires the axiom of choice.

Scott, Dana, *A proof of the independence of the continuum hypothesis*, Math. Systems Theory, 1 (1967) 89-111.

>A lucid exposition in which the author describes his proof of Cohen's theorem in language that is accessible to the mathematician who is not well-versed in mathematical logic. Winner of the LeRoy P. Steele prize for distinguished exposition of outstanding research.

Zlot, William Leonard, *The principle of choice in pre-axiomatic set theory*, Scripta Math., 25 (1960) 105-123.

>Historical survey of pre-Cohen and pre-independence results vintage.

2.1.3 RESEARCH

Keisler, H. Jerome and Tarski, Alfred, *From accessible to inaccessible cardinals*, Fund. Math., 53 (1963) 225-308.

>Major survey with long bibliography; requires familiarity with advanced techniques of mathematical logic.

Shoenfield, J.R., *Martin's axiom*, Amer. Math. Monthly, 82 (1975) 610-617.

The Gödel-Cohen results on the continuum hypothesis suggest (demand?) the development of a system in which there exists an intermediate cardinal. For example, is the union of a collection of sets of Lebesgue measure zero still of measure zero when the collection is of intermediate cardinality? Martin's axiom answers questions of this type. Among the implications of this new axiom for set theory are many of the consequences of the continuum hypothesis, together with important relations to Souslin's hypothesis.

Silver, Jack H., *The bearing of large cardinals on constructibility*, in M.D. Morley, Studies in Model Theory, M.A.A., 1973, pp. 158-182.

RELATED REFERENCES

Bing, R.H., *Challenging conjectures*, Amer. Math. Monthly, 74 (1967) Suppl. pp. 56-64. [5.5.2]

Blumenthal, Leonard M., *"A paradox, a paradox, a most ingenious paradox,"* Amer. Math. Monthly, 47 (1940) 346-353. [4.3.1]

Crossley, John N., et al., What is Mathematical Logic?, Oxford U. Pr., 1972. [2.6.2]

Dauben, Joseph W., *Denumerability and dimension: the origins of Georg Cantor's theory of sets*, Rete, 2 (1974) 105-133. [1.2.2]

Dauben, Joseph W., *The invariance of dimension: problems in the early development of set theory and topology*, Historia Math., 2 (1975) 273-288. [1.2.2]

Dauben, Joseph W., *The trigonometric background to Georg Cantor's theory of sets*, Arch. Hist. Exact Sci., 7 (1971) 181-216. [1.2.2]

Grattan-Guinness, Ivor, *Towards a biography of George Cantor*, Annals of Science, 27 (1971) 345-391. [1.3.0]

Mostowski, Andrezej, Thirty Years of Foundational Studies, Barnes & Noble, 1966. [1.2.2]

Rudin, Mary Ellen, *Souslin's conjecture*, Amer. Math. Monthly, 76 (1969) 1113-1119. [5.5.2]

Steen, Lynn Arthur, *New models of the real-number line*, Scientific American, 225 (August 1971) 92-99, 120. [2.4.1]

Ullian, Joseph S., *Is any set theory true?*, Phil. Sci., 36 (1969) 271-279. [2.2.2]

Wilder, Raymond L., *The role of the axiomatic method*, Amer. Math. Monthly, 74 (1967) 115-127; also in Math. Teaching, 41 (1967) 32-40. [1.5.0]

2.2 PHILOSOPHY OF MATHEMATICS

2.2.0 GENERAL

Agazzi, Evandro, *The rise of the foundational research in mathematics*, Synthese, 27 (1974) 7-26.
  Informal historical account of investigations in the foundations of mathematics.

Barker, Stephen F., *Realism as a philosophy of mathematics*, in J.J. Bulloff, T.C. Holyoke and S.W. Hahn, Foundations of Mathematics, Springer-Verlag, 1969, pp. 1-9.
  Gödel's viewpoint on the philosophy of mathematics.

Baum, Robert J., Philosophy and Mathematics, Freeman Cooper, 1973.

Davis, Chandler, *Materialist mathematics*, in R.S. Cohen and M.W. Wartofsky, Boston Studies in the Philosophy of Science, Humanities Pr., 1974, V. 15.
  Philosophy of mathematics from a dialectical materialist viewpoint.

Goodstein, R.L., *Empiricism in mathematics*, Dialectica, 23 (1969) 50-57.

Luchins, Edith H. and Luchins, Abraham S., *Logicism*, Scripta Math., 27 (1965) 223-243.
  Historical overview of logicism and summary of criticisms.

Pollock, John L., *Mathematical proof*, Amer. Phil. Quarterly, 4 (1967) 238-244.
  Argues that mathematical truths are discovered, not created.

Priest, Graham, *A bedside reader's guide to the conventionalist philosophy of mathematics*, in J. Bell, et al., Proc. Bertrand Russell Memorial Logic Conference, Leeds, 1973, pp. 115-132.
  Analyzes succinctly the leading philosophies of mathematics (Platonism, constructivism, logicism, formalism, empiricism) in terms of their answers to the following questions: What is (pure) mathematics? Why are its truths true? Why do such truths appear necessary and inviolable? How is it we come to know them? Why can they be applied in practical matters? What exactly are the objects of mathematics (such as numbers, functions, groups, models, etc.)? In what sense do they exist? If they don't exist, why is it that we have such a strong impression that they do? The conclusion that none of the traditional philosophies affords

satisfactory answers to these questions leads the author to formulate and extend the conventionalism of Carnap, Ayer, and the later Wittgenstein. "Conventionalism asserts that a truth of mathematics or logic is true in virtue of the meanings of its constituent words," and much of the power of explanation of conventionalism derives from its close identification of mathematics with language.

Putnam, Hilary, *Mathematics without foundations*, J. Phil., 64 (1967) 5-22.

Quine, Willard Van Orman, *The foundations of mathematics*, Scientific American, 211 (September 1964) 112-127, 269; also in M. Kline, Mathematics in the Modern World, Freeman, 1968, pp. 191-199, 399.

Robinson, Abraham, *Formalism 64*, in Y. Bar-Hillel, Logic, Methodology and Philosophy of Science, North-Holland, 1965, pp. 228-248.

Robinson, Abraham, *From a formalist's point of view*, Dialectica, 23 (1969) 45-49.

Rogers, Robert, *Mathematical and philosophical analyses*, Phil. Sci., 31 (1964) 255-264.

Sawyer, W.W., *A reflection on foundations of mathematics -- mathematicians regarded as biological specimens*, Philosophia Mathematica, 1 (1964) 5-32.

> Rambling but stimulating essay on the empirical and intuitionistic foundation of finite and infinite numbers.

2.2.1 ELEMENTARY

Bernays, Paul, *On Platonism in mathematics*, in P. Benacerraf and H. Putnam, Philosophy of Mathematics, Prentice-Hall, 1964, pp. 274-286.

Davis, Philip J., *Fidelity in mathematical discourse: is one and one really two?*, Amer. Math. Monthly, 79 (1972) 252-263.

> Controversial view of mathematical truth as inherently probabilistic.

Goodstein, R.L., *Existence in mathematics*, Compositio Math., 20 (1968) 70-82.

> Existence of numbers is equivalent to existence of functions. Written for both mathematicians and philosophers, it requires only modest mathematical background.

Grünbaum, Adolf, *Geometry, chronometry and empiricism*, in H. Feigel and G. Maxwell, <u>Minnesota Studies in the Philosophy of Science</u>, V. 3, U. Minn. Pr., 1962, pp. 405-526.

Grünbaum, Adolf, *Reply to Hilary Putnam's 'An examination of Grünbaum's philosophy of geometry'*, in R.S. Cohen and M.W. Wartofsky, <u>Boston Studies in the Philosophy of Science</u>, V. 5, D. Reidel, 1969, pp. 1-150.

Kleene, Stephen C. and Feferman, Solomon, *Foundations of mathematics*, <u>Encyclopaedia Britannica</u>, 15th ed., 1974, Macropaedia V. 11, pp. 630-639.

> The article includes a thorough discussion of the controversies occuring in this discipline since 1900, and a brief review of more recent trends.

2.2.2 ADVANCED

Bourbaki, Nicolas, *Foundations of mathematics for the working mathematician*, <u>J. Symbolic Logic</u>, 14 (1949) 1-8.

Casari, Ettore, *Axiomatical and set-theoretical thinking*, <u>Synthese</u>, 27 (1974) 49-61.

> Discusses two different ways of thinking in mathematics: set-theoretical reductionism and the arithmetization of analysis, vs. the axiomatic method; concludes that "set-theoretical thinking is really capable of absorbing completely--and perhaps of strengthening far beyond its natural limits--the modern abstract axiomatic method... ."

Cauman, Leigh S., *On indirect proof*, <u>Scripta Math.</u>, 28 (1968) 101-115.

> Exploration of the logical legitimacy of indirect proof.

Fraenkel, Abraham A., *The recent controversies about the foundations of mathematics*, <u>Scripta Math.</u>, 13 (1947) 17-36.

> Concise survey of problem of continuum, discussing intuitionism, formalism, logicism and Platonism. Fraenkel reacts pessimistically to Kronecker, Brouwer and Weyl's attempts to articulate intuitionistic mathematics, and concludes that it is not possible to construct the continuum.

Henkin, Leon, *Mathematical foundations for mathematics*, <u>Amer. Math. Monthly</u>, 78 (1971) 463-487.

> Three competitive views on the nature of the foundations of mathematics--intuitionism, formalism, and logicism--are associated with

the names Brouwer, Hilbert, and Russell. Using, respectively, "constructive", "algebraic", and "set-theoretic" as a related but different classification (which is complementary rather than competitive), the author surveys the current status of mathematical logic. An excellent first paper for anyone interested in learning more about this field. Winner of the Lester R. Ford award for expository writing.

Ullian, Joseph S., *Is any set theory true?*, Phil. Sci., 36 (1969) 271-279.

RELATED REFERENCES

Birnbaum, Allan, *On the foundations of statistical inference*, J. Amer. Stat. Assoc., 57 (1962) 269-326. [6.2.3]

Black, Max, *The elusiveness of sets*, Review of Metaphysics, 24 (1971) 614-636. [2.1.1]

Bronowski, Jacob, *The logic of the mind*, Amer. Scientist, 54 (1966) 1-14. [1.4.0]

Lakatos, Imre, *Proofs and refutations*, Brit. J. Phil. Science, 14 (1963-64) 1-25, 120-139, 221-245, 296-342. [1.5.2]

Monk, J. Donald, *On the foundations of set theory*, Amer. Math. Monthly, 77 (1970) 703-711. [2.1.2]

Rosenkrantz, R.D., *The significance test controversy*, Synthese, 26 (1973) 304-321. [6.2.2]

Weyl, Hermann, *Insight and reflection*, Studia Philosophica, 15 (1955); also in T.L. Saaty and F.J. Weyl, The Spirit and Uses of the Mathematical Sciences, McGraw-Hill, 1969, pp. 281-301. [1.5.0]

2.3 INTUITIONISM

2.3.0 GENERAL

Fraenkel, Abraham A., *On the crisis of the principle of the excluded middle*, Scripta Math., 17 (1951) 5-16.

Heyting, Arend, *Disputation*, in P. Benacerraf and H. Putnam, Philosophy of Mathematics, Prentice-Hall, 1964, pp. 55-65.
   Famous dialogue in which the views of intuitionism are aired.

Heyting, Arend, *The intuitionist foundations of mathematics*, in P. Benacerraf and H. Putnam, Philosophy of Mathematics, Prentice-Hall, 1964, pp. 42-49.

2.3.1 ELEMENTARY

Brouwer, L.E.J., *Historical background, principles, and methods of intuitionism*, South African J. Sci., 49 (1952-53) 139-146.

Brouwer, L.E.J., *Intuitionism and formalism*, Bull. Amer. Math. Soc., 20 (1913) 81-96; also in P. Benacerraf and H. Putnam, Philosophy of Mathematics, Prentice-Hall, 1964, pp. 66-77.
   Both philosophical and mathematical comparisons, with special emphasis on formalist axioms for set theory and their intuitionistic counterparts.

Heyting, Arend, *Intuitionistic views on the nature of mathematics*, Synthese, 27 (1974) 79-91.
   "Intuitionism is not a philosophical system on the same level with realism, idealism, or existentialism. The only philosophical thesis of mathematical intuitionism is that no philosophy is needed to understand mathematics." Various misconceptions about intuitionists' views on arithmetic, the continuum, set theory, logic, and language are described and set straight.

Lorenzen, P., *Constructive mathematics as a philosophical problem*, Compositio Math., 20 (1968) 133-142.

Myhill, John, *What is a real number?*, Amer. Math. Monthly, 79 (1972) 748-754.
   An intuitionistic (i.e., constructive) interpretation, replete with interesting pathological examples.

Stolzenberg, Gabriel, *Review of Errett Bishop's Foundations of Constructive Analysis*, Bull. Amer. Math. Soc., 76 (1970) 301-323.

> Lengthy review mixed with extensive historical exposition of intuitionistic mathematics. Stolzenberg argues that Bishop's results refute much of the criticism levelled at intuitionism in the early part of the 20th century.

2.3.2 ADVANCED

Aberth, Oliver, *Analysis in the computable number field*, J. Assoc. Comp. Mach., 15 (1968) 275-299.

> Surprising results appear to contradict fundamental theorems in analysis of the reals.

Posy, Carl J., *Brouwer's constructivism*, Synthese, 27 (1974) 125-159.

RELATED REFERENCES

Cauman, Leigh S., *On indirect proof*, Scripta Math., 28 (1968) 101-115. [2.2.2]

Fraenkel, Abraham A., *The recent controversies about the foundations of mathematics*, Scripta Math., 13 (1947) 17-36. [2.2.2]

Pierpont, James, *Mathematical rigor, past and present*, Bull. Amer. Math. Soc., 34 (1928) 23-53. [1.2.1]

Sawyer, W.W., *A reflection on foundations of mathematics--mathematicians regarded as biological specimens*, Philosophia Mathematica, 1 (1964) 5-32. [2.2.0]

Weyl, Hermann, *Mathematics and logic*, Amer. Math. Monthly, 53 (1946) 2-13. [2.6.1]

2.4 NONSTANDARD ANALYSIS

2.4.0 GENERAL

Robinson, Abraham, *Numbers--what are they and what are they good for?*, Yale Scientific, (May 1973) 14-16.

Steen, Lynn Arthur, *The metamathematical world of model theory*, Science News, 107 (1975) 108-111.

Sullivan, Walter, *New form of math said to resolve old paradoxes of numbering system*, New York Times (February 15)1975, 22.
   A very general account of Robinson's nonstandard analysis.

2.4.1 ELEMENTARY

Davis, Martin and Hersh, Reuben, *Nonstandard analysis*, Scientific American, 226 (June 1972) 78-86, 136.
   Well-written exposition of this fascinating subject, buttressed with interesting historical remarks.

Levitz, Hilbert, *Non-standard analysis: an exposition*, L'Enseignement Math., 20 (1974) 9-32.

Lightstone, A.H., *Infinitesimals*, Amer. Math. Monthly, 79 (1972) 242-251.
   "The goal of this article is to enliven Abraham Robinson's concept of an infinitesimal by exhibiting infinitesimals in a simple and direct manner." The presentation is via the compactness theorem of mathematical logic rather than by ultrafilters.

Robinson, Abraham, *Some thoughts on the history of mathematics*, Compositio Math., 20 (1968) 188-193.

Robinson, Abraham, *The metaphysics of the calculus*, in J. Hintikka, The Philosophy of Mathematics, Oxford U. Pr., 1969, pp. 153-163; also in I. Lakatos, Problems in the Philosophy of Mathematics, North-Holland, 1967, pp. 28-40.
   An exposition of the history and logical foundations of calculus with an emphasis on the nature of infinitary motions. Requires and rewards a careful reading.

Steen, Lynn Arthur, *New models of the real-number line*, Scientific American, 225 (August 1971) 92-99, 120.

> A broad description of two new "competitors" of the real number field which were first introduced by Abraham Robinson and Dana Scott. Robinson's model gives a theoretical approach to the notion of infinitesimal, while Scott's gives concrete formulation to Paul Cohen's results on the independence of the continuum hypothesis.

2.4.2 ADVANCED

Lightstone, A.H., *Infinitesimals and integration*, Math. Magazine, 46 (1973) 20-30.

Luxemburg, W.A.J., *What is nonstandard analysis?*, Amer. Math. Monthly, 80 (1973) Suppl. pp. 38-67.

> A "mainly expository" introduction using a formal language in which constants range over sets and functions. Includes surprising derivation of Euler's infinite product sine formula. Good bibliography.

Robinson, Abraham, *Function theory on some nonarchimedian fields*, Amer. Math. Monthly, 80 (1973) Suppl. pp. 87-109.

Robinson, Abraham, *Standard and nonstandard number systems*, Nieuw Archief Wiskunde, 21 (1973) 115-133.

Van Osdol, D.H., *Truth with respect to an ultrafilter or how to make intuition rigorous*, Amer. Math. Monthly, 79 (1972) 355-363.

2.4.3 RESEARCH

Bernstein, Allan R., *Non-standard analysis*, in M.D. Morley, Studies in Model Theory, M.A.A., 1973, pp. 35-58.

> In 1963, P.R. Halmos and K.T. Smith asked: If T is a bounded linear operator on a Hilbert space such that T^2 is compact, does T have a proper invariant subspace? The article presents the answer in the affirmative, as solved by A.R. Bernstein and A. Robinson in 1966.

Luxemburg, W.A.J., *A general theory of monads*, in W.A.J. Luxemburg, Applications of Model Theory to Algebra, Analysis, and Probability, Holt, Rinehart and Winston, 1969, pp. 18-86.

RELATED REFERENCES

Meisters, G.H. and Monk, J. Donald, *Construction of the reals via ultrapowers*, Rocky Mountain J. Math., 3 (1973) 141-158. [4.3.3]

Robinson, Abraham, *Between logic and mathematics*, I.C.S.U. Rev. World Sci., 6 (1964) 218-226. [2.6.1]

Robinson, Abraham, *Formalism 64*, in Y. Bar-Hillel, Logic, Methodology and Philosophy of Science, North-Holland, 1965, pp. 228-248. [2.2.0]

Robinson, Abraham, *From a formalists' point of view*, Dialectica, 23 (1969) 45-49. [2.2.0]

Rubin, Jean E., *The compactness theorem in mathematical logic*, Math. Magazine, 46 (1973) 261-265. [2.6.2]

2.5 DECIDABILITY

2.5.0 GENERAL

Bernhard, Robert, *Crisis in math--is there 'universal truth'?*, Scientific Research, (October 14, 1968) 47-56.

Chaitin, Gregory J., *Randomness and mathematical proof*, Scientific American, 232 (May 1975) 47-52, 122.
> "Although randomness can be precisely defined and can even be measured, a given number cannot be proved to be random. This enigma establishes a limit to what is possible in mathematics." The article relates this limitation to Gödel's incompleteness results.

DeSua, Frank C., *Metamathematics: a non-technical exposition*, Amer. Scientist, 42 (1954) 488-495.
> Exposition of Gödel's work.

Nagel, Ernest and Newman, James R., *Gödel's proof*, Scientific American, 194 (June 1956) 71-86, 168, 170; also in M. Kline, Mathematics in the Modern World, Freeman, 1968, pp. 221-230, 400; in J.R. Newman, The World of Mathematics, V. 3, 1956, pp. 1668-1695; also published by New York U. Pr., 1958.
> A famous essay explicating the most important result of mathematical logic in the past millenium. The details are omitted, yet even a cursory understanding requires a very careful reading. Provides an excellent illustration of the role of the axiomatic method in modern mathematics. The philosophical implications of Gödel's theorem are discussed at some length.

Smullyan, Raymond M., *Review of Kurt Gödel's On Formally Undecidable Propositions of Principia Mathematica and Related Systems*, Amer. Math. Monthly, 73 (1966) 319-322.
> A careful essay on Kurt Gödel's undecidability results.

Steen, Lynn Arthur, *Foundations of mathematics: unsolvable problems*, Science, 189 (1975) 209-210.

2.5.1 ELEMENTARY

DeLong, Howard, *Unsolved problems in arithmetic*, Scientific American, 224 (March 1971) 50-60, 124.
> "It is the nature of arithmetic that it can pose more problems than it can solve. Indeed, one of the triumphs of mathematical

logic is the demonstration that there are problems that can never be solved." An essay providing interesting examples to illustrate this thesis.

DeSua, Frank C., *Consistency and completeness--a résumé*, Amer. Math. Monthly, 63 (1956) 295-305.

Historical survey providing background for Gödel's work.

Kempner, A.J., *Remarks on 'unsolvable' problems*, Amer. Math. Monthly, 43 (1936) 467-473.

Discusses types of unsolvability with examples from number theory, e.g., decimal expansion of pi, Goldbach's conjecture. Excellent source of ideas of significance for modern undecidability results. Easy reading.

Kleene, Stephen C., *The new logic*, Amer. Scientist, 57 (1969) 333-347.

Gödel results and computability.

Naps, Thomas L., *Arithmetical formalism--Hilbert's proof theory and Gödel's proof*, J. Undergraduate Math., 1 (1969) 111-132.

2.5.2 ADVANCED

Ershov, Y.L., et al., *Elementary theories*, Russian Math. Surveys, 20:4 (1965) 35-105.

Comprehensive survey, with appendix charts listing all known decidable and undecidable theories.

Jones, James P., *Recursive undecidability--an exposition*, Amer. Math. Monthly, 81 (1974) 724-738.

A simple treatment, based on combinatorial rather than formal arithmetization arguments, of computability, unsolvability, and variations on the halting problem.

Meserve, Bruce E., *Decision methods for elementary algebra*, Amer. Math. Monthly, 62 (1955) 1-8.

Exposition of Tarski's methods for deciding truth of sentences of elementary algebra.

Rosser, J. Barkley, *An informal exposition of proofs of Gödel's theorems and Church's theorem*, J. Symbolic Logic, 4 (1939) 53-60.

RELATED REFERENCES

Boone, William W., *The word problem*, Annals of Math., 70 (1959) 207-265. [3.5.3]

Crossley, John N., et al., *What is Mathematical Logic?*, Oxford U. Pr., 1972. [2.6.2]

Davis, Martin, *Hilbert's tenth problem is unsolvable*, Amer. Math. Monthly, 80 (1973) 233-269. [3.3.3]

Davis, Martin and Hersh, Reuben, *Hilbert's 10th problem*, Scientific American, 229 (November 1973) 84-91, 136. [3.3.1]

Gödel, Kurt, *What is Cantor's continuum problem?*, Amer. Math. Monthly, 54 (1947) 515-525; also in P. Benacerraf and H. Putnam, Philosophy of Mathematics, Prentice-Hall, 1964, pp. 258-273. [2.1.2]

Hanson, Norwood Russell, *Number theory and physical theory: an analogy*, in R.S. Cohen and M.W. Wartofsky, Boston Studies in the Philosophy of Science, V. 2, Humanities Pr., 1965, pp. 93-119. [7.1.1]

Hilbert, David, *On the infinite*, in P. Benacerraf and H. Putnam, Philosophy of Mathematics, Prentice-Hall, 1964, pp. 134-151. [2.1.1]

Lambek, Joachim, *The mathematics of sentence structures*, Amer. Math. Monthly, 65 (1958) 154-170. [7.6.2]

Robinson, Julia, *Diophantine decision problems*, in W.J. LeVeque, Studies in Number Theory, M.A.A., 1969, pp. 76-116. [3.3.3]

Rogers, Hartley R., Jr., *The present state of Turing machine computability*, SIAM J. Appl. Math., 7 (1959) 114-130. [6.5.2]

Shepherdson, J.C. and Sturgis, H.E., *Computability of recursive functions*, J. Assoc. Comp. Mach., 10 (1963) 217-255. [6.5.2]

Trakhtenbrot, B.A., *Algorithms*, in Z.W. Pylyshyn, Perspectives on the Computer Revolution, Prentice-Hall, 1970, pp. 69-86. [6.5.1]

2.6 LOGIC

2.6.0 GENERAL

Gardner, Martin, <u>Logic Machines, Diagrams and Boolean Algebras</u>, Dover, 1968.

Henkin, Leon, *Are logic and mathematics identical?*, <u>Science</u>, 138 (1962) 788-794.
  Very informative brief history of modern logic and its relation to mathematics. Winner of the Chauvenet Prize for expository writing in mathematics.

Kneebone, G.T., *Logic*, in N.J. Hardiman, <u>Exploring University Mathematics</u>, V. 3, Pergamon, 1969, pp. 68-79.

Quine, Willard Van Orman, *Paradox*, <u>Scientific American</u>, 206 (April 1962) 84-96, 193; also in M. Kline, <u>Mathematics in the Modern World</u>, Freeman, 1968, pp. 200-208, 399.
  The author relates interesting paradoxes with even more interesting explanations and classifies paradoxes as veridical, falsidical, or antinomic.

Tarski, Alfred, *Truth and proof*, <u>Scientific American</u>, 220 (June 1969) 63-77, 144.

2.6.1 ELEMENTARY

Black, Max, *Induction*, <u>Encyclopedia Americana</u>, 1976, V. 15, p. 100.

Blumenthal, Leonard M., *Logic*, <u>McGraw-Hill Encyclopedia of Science and Technology</u>, 1960, V. 7, pp. 578-582.

Buck, R.C., *Mathematical induction and recursive definitions*, <u>Amer. Math. Monthly</u>, 70 (1963) 128-135.

Henkin, Leon, *On mathematical induction*, <u>Amer. Math. Monthly</u>, 67 (1960) 323-338.

Randall, C.H. and Foulis, D.J., *An approach to empirical logic*, <u>Amer. Math. Monthly</u>, 77 (1970) 363-374.
  "The aim of an empirical science is to order, explain, and predict the observable consequences of certain physical operations. The mathematical apparatus for such an empirical science, therefore,

ought to be based on a formal 'empirical logic' capable of describing these events. Such logics, in turn, ought to be erected on a refinement of the conventional notion of a physical operation. A rigorous development, based on this desideratum, leads to a nonclassical symbolic logic that bears a striking resemblance to the so-called logic of quantum mechanics."

Robinson, Abraham, *Between logic and mathematics*, **I.C.S.U. Rev. World Sci.**, 6 (1964) 218-226.

Rogers, Hartley R., Jr., *An example in mathematical logic*, **Amer. Math. Monthly**, 70 (1963) 929-945.

An elimination-of-quantifiers decision procedure in mathematical logic is used to bring up to date "Leibnitz's dream", which was "that man should discover: (a) a precisely definable universal symbolism for making statements of science (Leibnitz called this a *characteristica universalis*); and (b) an algorithm which, when applied to the symbols of any formula of the characteristica universalis would determine whether or not that formula were true as a statement of science... ."

Weyl, Hermann, *Mathematics and logic*, **Amer. Math. Monthly**, 53 (1946) 2-13.

A personal survey, stressing both history and opinion.

2.6.2 ADVANCED

Crossley, John N., et al., **What is Mathematical Logic?**, Oxford U. Pr., 1972.

Lectures in an admirably informal style, treating in turn the history of mathematical logic, the completeness of the predicate calculus, model theory, Turing machines and recursive functions, Gödel's incompleteness theorems, and set theory and the continuum hypothesis.

Keisler, H. Jerome, *Model theory*, in **Actes Cong. Inter. Math.** (1970), V. 1, Gauthier-Villars, 1971, pp. 141-150.

Kreisel, Georg, *Mathematical logic*, in T.L. Saaty, **Lectures on Modern Mathematics**, V. 3, Wiley, 1965, pp. 95-195.

Putnam, Hilary, *Recursive functions and hierarchies*, **Amer. Math. Monthly**, 80 (1973) Suppl. pp. 68-86.

Rasiowa, Helena, *Post algebras as a semantic foundation of m-valued logics*, in A. Daigneault, **Studies in Algebraic Logic**, M.A.A., 1974, pp. 92-142.

The connections between Post algebras and propositional and first order m-valued logics are described. Extensive bibliography.

Robinson, Abraham, *Model theory as a framework for algebra*, in M.D. Morley, Studies in Model Theory, M.A.A., 1973, pp. 134-157.

> The article shows "how certain basic facts and notions of Algebra, for example the notion of an algebraically closed field, can be placed and generalized within the framework of Model Theory."

Rubin, Jean E., *The compactness theorem in mathematical logic*, Math. Magazine, 46 (1973) 261-265.

Vaught, Robert L., *Models of complete theories*, Bull. Amer. Math. Soc., 69 (1963) 299-313.

> Brisk survey of satisfaction, truth, model. Several examples.

2.6.3 RESEARCH

Barwise, Jon, *Back and forth through infinitary logic*, in M.D. Morley, Studies in Model Theory, M.A.A., 1973, pp. 5-34.

> Discussion of the significance of the "back and forth" technique introduced by Cantor; with applications.

Chang, C.C., *What's so special about saturated models?*, in M.D. Morley, Studies in Model Theory, M.A.A., 1973, pp. 59-95.

> Survey of saturated and special models with applications to preservation theorems, definability and interpolation, and two-cardinal systems.

Craig, William, *Unification and abstraction in algebraic logic*, in A. Daigneault, Studies in Algebraic Logic, M.A.A., 1974, pp. 6-57.

> An axiomatization of the theory of polyadic algebras with equality which uses only transformations, Boolean operations and diagonal elements.

Crossley, John N., *Recursive equivalence*, Bull. London Math. Soc., 2 (1970) 129-151.

> "Classical set theory and the theory of computable (recursive) functions are the basic areas between which the theory of recursive equivalence forms a bridge. ...[It] also has connections with other parts of logic. In particular Tarski's work on cardinal and ordinal algebras profoundly influenced the original development of the theories of recursive equivalence types and constructive order types." The author surveys recent developments in this field.

Halmos, Paul R., *The basic concepts of algebraic logic*, Amer. Math. Monthly, 63 (1956) 363-387.

Keisler, H. Jerome, *A survey of ultraproducts*, in Y. Bar-Hillel, Logic, Methodology and Philosophy of Science, North-Holland, 1965, pp. 112-126.

Keisler, H. Jerome, *Forcing and the omitting types theorem*, in M.D. Morley, Studies in Model Theory, M.A.A., 1973, pp. 96-133.

Monk, J. Donald, *Connections between combinatorial theory and algebraic logic*, in A. Daigneault, Studies in Algebraic Logic, M.A.A., 1974, pp. 58-91.

> Cylindric algebras are associated in turn with loops, projective geometries, and certain graphs. Results about the latter structures are obtained by considering the representability of the associated cylindric algebras, e.g., a loop is a group if and only if the algebra is representable.

Reyes, Gonzalo E., *From sheaves to logic*, in A. Daigneault, Studies in Algebraic Logic, M.A.A., 1974, pp. 143-204.

> A study of "the 'geometric' and logical aspects of a category of sheaves."

Vaught, Robert L., *Some aspects of the theory of models*, Amer. Math. Monthly, 80 (1973) Suppl. pp. 3-37.

> A brief account of the basic notions of model theory and a proof of Gödel's compactness theorem precede more advanced "aspects."

RELATED REFERENCES

Bourbaki, Nicolas, *Foundations of mathematics for the working mathematician*, J. Symbolic Logic, 14 (1949) 1-8. [2.2.2]

Henkin, Leon, *Mathematical foundations for mathematics*, Amer. Math. Monthly, 78 (1971) 463-487. [2.2.2]

Kac, Mark and Ulam, Stanislaw M., Mathematics and Logic: Retrospect and Prospects, Frederick A. Praeger, 1968; The New American Library, 1969. [1.1.2]

Kasner, Edward and Newman, James R., *Paradox lost and paradox regained*, in J.R. Newman, The World of Mathematics, V. 3, Simon and Schuster, 1956, pp. 1936-1955. [2.1.0]

Lieber, Lillian R., <u>Mits, Wits, and Logic</u>, [1.7.0]
Norton, 1947; 1954; 1960.

Mostowski, Andrezej, <u>Thirty Years of Founda- [1.2.2]
tional Studies</u>, Barnes & Noble, 1966.

Robinson, J.A., *Theorem-proving on the compu- [6.5.2]
ter*, <u>J. Assoc. Comp. Mach.</u>, 10 (1963) 163-174.

Wilder, Raymond L., *The role of the axiomatic [1.5.0]
method*, <u>Amer. Math. Monthly</u>, 74 (1967) 115-
127; also in <u>Math. Teaching</u>, 41 (1967) 32-40.

3.1 Combinatorics

3.1.0 GENERAL

Greene, Francis A., *Combinations and permutations*, Encyclopedia Americana, 1976, V. 7, pp. 358-360.

3.1.1 ELEMENTARY

Grünbaum, Branko, *Polygons in arrangements generated by n points*, Math. Magazine, 46 (1973) 113-119.
> "In 1826 Jacob Steiner initiated the study of arrangement of lines in the plane, that is, of the various ways in which a finite set of lines may partition the plane." This article deals with a generalization of the following arrangement problem: "In a convex polygon all diagonals are drawn, decomposing the polygon into smaller ones. What is the greatest number of sides possible for one of the smaller polygons...?"

Hall, Marshall, Jr. and Knuth, Donald E., *Combinatorial analysis and computers*, Amer. Math. Monthly, 72 (1965) Suppl. pp. 21-28.
> Generation of sequences of combinatorial patterns, backtrack, Latin squares, projective planes and symmetric block designs.

Honsberger, Ross, Ingenuity in Mathematics, New Math. Libr., No. 23, Random House, 1970; M.A.A., 1975.

Honsberger, Ross, Mathematical Gems, M.A.A., 1973.
> The "gems" are contained in 13 essays on elementary combinatorics, number theory, and geometry. The author hopes that the book will be "of special interest to high school mathematics teachers and to prospective teachers." It should appeal to a wider audience; for example, Chapter 7 describes a 1968 result of Grinberg on Hamiltonian circuits and how it leads to the best counterexample to Tait's 1880 conjecture of 4-color problem fame.

Kolata, Gina Bari, *Combinatorics: steps toward a unified theory*, Science, 183 (1974) 839-840, 883.

Niven, Ivan M., Mathematics of Choice--or How to Count Without Counting, New Math. Libr., No. 15, Random House, 1965; M.A.A., 1975.
> An exposition of combinatorial analysis at an advanced high school level, including a large problem selection.

Rota, Gian-Carlo, *Combinatorial analysis*, in <u>The Mathematical Sciences--A Collection of Essays for COSRIMS</u>, M.I.T., 1969, pp. 197-208.

Seven examples of challenging problems, among them the so-called Ising problem: given a rectangular grid made up of unit squares, each colored either red or blue, how many different color patterns are there, if the number of boundary edges between the red squares and the blue squares is prescribed? This question turns out to be equivalent to one in statistical mechanics, bearing on no less a matter than the explanation of the "macroscopic behavior of matter on the basis of known facts at the molecular or atomic levels."

3.1.2 ADVANCED

Alder, Henry L., *Partition identities--from Euler to the present*, <u>Amer. Math. Monthly</u>, 76 (1969) 733-746.

Winner of the Lester R. Ford award for expository writing.

Bender, Edward A., *Asymptotic methods in enumeration*, <u>SIAM Review</u>, 16 (1974) 485-515.

Harary, Frank and Moser, Leo, *The theory of round robin tournaments*, <u>Amer. Math. Monthly</u>, 73 (1966) 231-246.

Bibliography of 20 references.

Koehler, J.E., *Folding a strip of stamps*, <u>J. Combinatorial Theory</u>, 5 (1968) 135-152.

Definitive solution. Hard.

Rota, Gian-Carlo and Harper, L.H., *Matching theory: an introduction*, <u>Adv. in Prob.</u>, 1 (1971) 169-215.

Dilworth's theorem, the marriage theorem, and other basic results, plus applications (Birkhoff-von Neumann theorem) and significant extensions. A more sophisticated version of Rota's film "The Marriage Theorem", but done in the same style of absolute clarity.

Tucker, Alan, *Pólya's enumeration formula by example*, <u>Math. Magazine</u>, 47 (1974) 248-256.

RELATED REFERENCES

Assmus, E.F., Jr. and Mattson, H.F., Jr., [7.5.3] *Coding and combinatorics*, <u>SIAM Review</u>, 16 (1974) 349-388.

Bose, Raj C. and Grünbaum, Branko, *Combina-*   [5.2.2]
torics and combinatorial geometry, Encyclopaedia Britannica, 15th ed., 1974, Macropaedia V. 4, pp. 942-954.

Golomb, Solomon W. and Baumert, Leonard D.,   [6.5.2]
Backtrack programming, J. Assoc. Comp. Mach.,
12 (1965) 516-524.

Klee, Victor, *The Euler characteristic in*   [5.2.3]
combinatorial geometry, Amer. Math. Monthly,
70 (1963) 119-127.

Monk, J. Donald, *Connections between combina-*   [2.6.3]
torial theory and algebraic logic, in A.
Daigneault, Studies in Algebraic Logic,
M.A.A., 1974, pp. 58-91.

Seifert, Herbert and Threlfall, William, *Old*   [5.7.3]
and new results on knots, Canad. J. Math., 2
(1950) 1-15.

3.2 GRAPH THEORY

3.2.0 GENERAL

Franklin, Philip, *The four color problem*, Scripta Math., 6 (1939) 149-156, 197-210; excerpted in Galois Lectures, Scripta Mathematica, 1941, pp. 53-85.

  Lucid exposition of one of mathematics' most famous unsolved problems. Contains many historical references.

Harary, Frank, *Some historical and intuitive aspects of graph theory*, SIAM Review, 2 (1960) 123-131.

  Good survey and introduction. No proofs.

Ore, Oystein, Graphs and Their Uses, New Math. Libr., No. 10, Random House, 1963; M.A.A., 1975.

Saaty, Thomas L., *Remarks on the four color problem: the Kempe catastrophe*, Math. Magazine, 40 (1967) 31-36.

  The error in Kempe's attempt to prove the four color conjecture sheds light on the difficulty encountered in using the inductive approach to attack this famous problem.

3.2.1 ELEMENTARY

Berge, Claude, *Graph theory*, Amer. Math. Monthly, 71 (1964) 471-481.

  The author, whose 1962 book The Theory of Graphs and Its Applications is one of the standard accounts of the subject, gives here a quick overview of certain aspects of graph theory. He gives no proofs, but states some important basic theorems and explains their relevance to various problems. He begins with a review, with illustrative examples, of rudimentary definitions.

Coxeter, H.S. MacDonald, *Map coloring problems*, Scripta Math., 23 (1957) 11-25.

  Application to Möbius bands and Klein bottles.

Harary, Frank, *Graph theory*, McGraw-Hill Encyclopedia of Science and Technology, 1960, V. 6, pp. 253-256.

Harary, Frank, *On the history of the theory of graphs*, in F. Harary, New Directions in the Theory of Graphs, Academic Pr., 1973, pp. 1-17.

Minty, G.J., *On the axiomatic foundations of the theories of directed linear graphs, electrical networks and network-programming*, J. Math. and Mech., 15:3 (1966); also in D.R. Fulkerson, Studies in Graph Theory, Part II, M.A.A., 1975, pp. 246-300.

O'Neil, P.V., *Ulam's conjecture and graph reconstructions*, Amer. Math. Monthly, 77 (1970) 35-43.

> The author utilizes a conjecture of Ulam's as a vehicle to describe some elementary aspects of graph theory. Winner of the Lester R. Ford award for expository writing.

Tutte, W.T., *Map-coloring problems and chromatic polynomials*, Amer. Scientist, 62 (1974) 702-705.

Tutte, W.T., *Symmetrical graphs and coloring problems*, Scripta Math., 25 (1960) 305-316.

> Construction and mathematical significance of some very symmetrical graphs.

3.2.2 ADVANCED

Brualdi, R.A., *Transversal theory and graphs*, in D.R. Fulkerson, Studies in Graph Theory, Part I, M.A.A., 1975, pp. 23-88.

Saaty, Thomas L., *Thirteen colorful variations on Guthrie's four-color conjecture*, Amer. Math. Monthly, 79 (1972) 2-43.

> A rather extensive historical survey of the four-color problem and of closely related topics in the theory of graphs. Winner of the Lester R. Ford award for expository writing.

Turner, James and Kautz, William H., *A survey of progress in graph theory in the Soviet Union*, SIAM Review, 12 (1970) Suppl. pp. 1-68.

> A survey of virtually all Soviet work in graph theory and its applications (to June 1968), correlated with comparable Western work. The authors consider Soviet work to be improving rapidly, but nonetheless lagging behind corresponding Western work. They report that the strongest areas of Soviet graph theory are: (1) bounds on numerical indices associated with graphs, (2) properties of algebraic structures associated with graphs, and (3) operations on graphs.

Tutte, W.T., *Chromials*, in D.R. Fulkerson, Studies in Graph Theory, Part II, M.A.A., 1975, pp. 361-377.

Whitney, Hassler and Tutte, W.T., *Kempe chains and the four color problem*, <u>Utilitas Mathematica</u>, 2 (November 1972); also in D.R. Fulkerson, <u>Studies in Graph Theory</u>, Part II, M.A.A., 1975, pp. 378-413.

> In 1971 Y. Shimamoto proposed a proof of the four color problem which was based on mathematical reasoning plus a computer analysis. An inaccurate computer program led to wild rumors of success, as well as misgivings and skepticism. The authors were led to a careful analysis of the mathematical portion, and then to this account of (an approach to) the four color problem, starting from the beginning. "We have tried to clarify the theory for ourselves, and we dare to hope that we may thereby have clarified it for others."

Wilson, R.J., *An introduction to matroid theory*, <u>Amer. Math. Monthly</u>, 80 (1973) 500-525.

> In the 1930's Whitney "...noticed several similarities between the ideas of independence and rank in graph theory and those of linear independence and dimension in...vector spaces. [He introduced] the concept of a matroid to formalize these similarities." The author describes how the concept lay dormant until 1958, how that changed, and the present relevance of matroids to graph theory and to transversal theory. Winner of the Lester R. Ford award for expository writing.

3.2.3 RESEARCH

Berge, Claude, *Perfect graphs*, in D.R. Fulkerson, <u>Studies in Graph Theory</u>, Part I, M.A.A., 1975, pp. 1-22.

Grünbaum, Branko, *Polytopal graphs*, in D.R. Fulkerson, <u>Studies in Graph Theory</u>, Part II, M.A.A., 1975, pp. 201-224.

Hoffman, Alan J., *Eigenvalues of graphs*, in D.R. Fulkerson, <u>Studies in Graph Theory</u>, Part II, M.A.A., 1975, pp. 225-245.

> Interplay between properties of a graph and the spectrum (set of eigenvalues) of its adjacency matrix.

Nash-Williams, C. St.J.A., *Hamiltonian circuits*, in D.R. Fulkerson, <u>Studies in Graph Theory</u>, Part II, M.A.A., 1975, pp. 301-360.

RELATED REFERENCES

Dantzig, George B., *On the shortest route through a network*, <u>Management Science</u>, 6 (January 1960); also in D.R. Fulkerson, <u>Studies in Graph Theory</u>, Part I, M.A.A., 1975, pp. 89-93. [6.3.1]

Duffin, R.J., *Network models*, <u>SIAM-AMS Proc.</u>, 3 (1971) 65-91; an expanded version appears as *Electrical network models* in D.R. Fulkerson, <u>Studies in Graph Theory</u>, Part I, M.A.A., 1975, pp. 94-138. [7.1.2]

Fulkerson, D.R., *Flow networks and combinatorial operations research*, <u>Amer. Math. Monthly</u>, 73 (1966) 115-138; also in D.R. Fulkerson, <u>Studies in Graph Theory</u>, Part I, M.A.A., 1975, pp. 139-171. [6.3.1]

Gomory, Ralph E. and Hu, T.C., *Multi-terminal flows in a network*, <u>SIAM J. Appl. Math.</u>, 9 (1961) 551-570; also in D.R. Fulkerson, <u>Studies in Graph Theory</u>, Part I, M.A.A., 1975, pp. 172-199. [6.3.2]

Grossman, Israel and Magnus, Wilhelm, <u>Groups and Their Graphs</u>, New Math. Libr., No. 14, Random House, 1964; M.A.A., 1975. [3.5.0]

Haggett, Peter, *Network models in geography*, in R.J. Chorley and P. Haggett, <u>Models in Geography</u>, Methuen, 1967, pp. 609-668. [7.6.2]

Honsberger, Ross, <u>Mathematical Gems</u>, M.A.A., 1973. [3.1.1]

Lederberg, Joshua, *Topology of molecules*, in <u>The Mathematical Sciences--A Collection of Essays for COSRIMS</u>, M.I.T., 1969, pp. 37-51. [7.1.0]

Marimont, Rosalind B., *Applications of graphs and Boolean matrices to computer programming*, <u>SIAM Review</u>, 2 (1960) 259-268. [6.5.1]

Mech, William P., *Graphs of groups*, <u>J. Undergraduate Math.</u>, 1 (1969) 97-100; 2 (1970) 37-49. [3.5.2]

Rashevsky, N., *Topology and life--in search* [7.2.1]
of general mathematical principles in biology
and sociology, Bull. Math. Biophys., 16
(1954) 317-348.

Roberts, Fred S. and Brown, Thomas A., *Signed* [7.6.1]
diagraphs and the energy crisis, Amer. Math.
Monthly, 82 (1975) 577-594.

3.3 NUMBER THEORY

3.3.0 GENERAL

Davis, Philip J., The Lore of Large Numbers, New Math. Libr., No. 6, Random House, 1961; M.A.A., 1975.

Hoffer, William, *A magic ratio recurs throughout art and nature*, Smithsonian, 6 (December 1975) 110-124.

Niven, Ivan M., Numbers: Rational and Irrational, New Math. Libr., No. 1, Random House, 1961; M.A.A., 1975.

Olds, Carl Douglas, Continued Fractions, New Math. Libr., No. 9, Random House, 1963; M.A.A., 1975.

3.3.1 ELEMENTARY

Archibald, Ralph G., *Goldbach's theorem*, Scripta Math., 3 (1935) 44-50, 153-161.

Batchelder, P.M., *Waring's problem*, Amer. Math. Monthly, 43 (1936) 21-27.
   An introduction to the problem of representing positive integers as sums of like powers of other positive integers.

Davis, Martin and Hersh, Reuben, *Hilbert's 10th problem*, Scientific American, 229 (November 1973) 84-91, 136.
   Winner of the Chauvenet Prize for expository writing in mathematics.

Dickson, L.E., *The Waring problem and its generalizations*, Bull. Amer. Math. Soc., 42 (1936) 833-842.
   Historical outline of solution, without proofs.

Evans, Trevor, *Nonassociative number theory*, Amer. Math. Monthly, 64 (1957) 299-309.
   Algebra of exponents of the general element in a nonassociative linear algebra.

Jones, Burton W., *Theory of numbers*, Encyclopedia Americana, 1976, V. 20, pp. 538-541.

Mardzanisvili, K.K. and Postnikov, A.B., *Prime numbers*, in A.D. Aleksandrov, et al., Mathematics--Its Content, Methods and Meaning, V. 2, M.I.T., 1963, pp. 199-228.

Mordell, L.J., *Three Lectures on Fermat's Last Theorem*, Macmillan, 1921; reprinted in F. Klein, et al., *Famous Problems*, Chelsea, 1962.

Ore, Oystein, *Invitation to Number Theory*, New Math. Libr., No. 20, Random House, 1967; M.A.A., 1975.

Pólya, George, *Heuristic reasoning in the theory of numbers*, *Amer. Math. Monthly*, 66 (1959) 375-384.

Rademacher, Hans, *Number theory*, *McGraw-Hill Encyclopedia of Science and Technology*, 1960, V. 9, pp. 224-227.

Saaty, Thomas L. and Alexander, Joyce M., *Optimization and the geometry of numbers: packing and covering*, *SIAM Review*, 17 (1975) 475-519.

  A lively introduction to the geometry of numbers.

Sierpiński, W., *On some unsolved problems of arithmetic*, *Scripta Math.*, 25 (1960) 125-136.

  Are there three rationals, the sum as well as the product of which equals 1? Surprisingly, the answers to this and to other simply-formulated problems are still unknown.

Waugh, Frederick V. and Maxfield, Margaret W., *Side-and-diagonal numbers*, *Math. Magazine*, 40 (1967) 74-83.

  Winner of the Lester R. Ford award for expository writing.

Weil, André, *Two lectures on number theory, past and present*, *L'Enseignement Math.*, 20 (1974) 87-110.

  Fascinating historical talks emphasizing the continuity of number theory for the last 300 years.

3.3.2 ADVANCED

Archibald, Ralph G., *Waring's problem: squares*, *Scripta Math.*, 7 (1940) 33-48.

Billingsley, Patrick, *On the central limit theorem for the prime divisor function*, *Amer. Math. Monthly*, 76 (1969) 132-139.

  Hardy and Ramanujan proved that "practically all" integers m have about log log m prime divisors--"practically all" and "about" are defined in this article. Thus, e.g., a "typical" integer in the vicinity of 100,000,000 will have only about three prime divisors. After briefly reviewing the history of the subject, the author presents a simplified proof of a key theorem by emphasizing its probabilistic nature.

[3.3] 70 Algebra:

Billingsley, Patrick, *Prime numbers and Brownian motion*, Amer. Math. Monthly, 80 (1973) 1099-1115.

> The author begins with a quick review of (mathematical) Brownian motion and Wiener measure, wherever possible motivating results intuitively rather than describing proofs. He then describes one dimensional random walks, in order to set the stage for the interaction of probability theory with number theory. To oversimplify, the entreé into number theory is gained by starting with the fact that the (approximate) probabilities that a large random number is divisible by p, q, or pq are respectively 1/p, 1/q, and 1/pq, so that the probabilities for primes p and q are independent. Winner of the Lester R. Ford award for expository writing.

Ellison, W.J., *Waring's problem*, Amer. Math. Monthly, 78 (1971) 10-36.

> Winner of the Lester R. Ford award for expository writing and of the LeRoy P. Steele prize for distinguished exposition of outstanding research.

Erdös, Paul, *Some recent advances and current problems in number theory*, in T.L. Saaty, Lectures on Modern Mathematics, V. 3, Wiley, 1965, pp. 196-244.

Ferguson, Rolfe P., *On Fermat's last theorem*, J. Undergraduate Math., 6 (1974) 1-14, 85-97; 7 (1975) 35-45.

> Detailed historical survey.

Goldstein, L.J., *A history of the prime number theorem*, Amer. Math. Monthly, 80 (1973) 599-615; Correction, 80 (1973) 1115.

> The author traces the history of the prime number theorem, with emphasis on the contributions of Gauss, Dirichlet, and Riemann, including the role of the Riemann Hypothesis in guiding the way.

Hardy, G.H., *An introduction to the theory of numbers*, Bull. Amer. Math. Soc., 35 (1929) 778-818.

> Winner of the Chauvenet Prize for expository writing in mathematics.

Hille, Einar, *Gelfond's solution of Hilbert's seventh problem*, Amer. Math. Monthly, 49 (1942) 654-661.

> Detailed exposition. Irrational and transcendental numbers.

James, R.D., *Recent progress in the Goldbach problem*, Bull. Amer. Math. Soc., 55 (1949) 246-260.

Lehmer, D.H., *Computer technology applied to the theory of numbers*, in W.J. LeVeque, Studies in Number Theory,

M.A.A., 1969, pp. 117-151.

> "We say that [an]...algorithm is of order f(n) in the parameter n if the time of execution is less than a constant multiple of f(n)." The author discusses basic algorithms of various orders (e.g., greatest common divisor, determination of primitive root, tests for primality) with comments on the inadequacy of standard machine languages for number theoretic computations.

LeVeque, William J., *A brief survey of diophantine equations*, in W.J. LeVeque, Studies in Number Theory, M.A.A., 1969, pp. 4-24.

> A lucid and non-technical overview beginning with solutions in rational numbers or integers of polynomial equations f(x,y) = 0. Includes brief accounts of basic results due to Hilbert and Hurwitz, Poincaré, and Thue, of later important results by Mordell, Weil, and Siegel, of Waring-type problems, and of various results related to the Artin conjecture and the Ax-Kochen theorem. Shows how the analysis of p-adic solutions can be used as a tool in diophantine problems.

Levinson, Norman, *A motivated account of an elementary proof of the prime number theorem*, Amer. Math. Monthly, 76 (1969) 225-245.

> This paper is designed to make as accessible as possible the author's simplified proof of the famed prime number theorem. Winner of the Lester R. Ford award and the Chauvenet Prize for expository writing in mathematics.

Olds, Carl Douglas, *The simple continued fraction expansion of e*, Amer. Math. Monthly, 77 (1970) 968-974.

> Winner of the Chauvenet Prize for expository writing in mathematics.

Robinson, Raphael M., *The converse of Fermat's theorem*, Amer. Math. Monthly, 64 (1957) 703-710.

Schmidt, Wolfgang M., *Approximation to algebraic numbers*, L'Enseignement Math., 17 (1971) 187-253.

> Survey lecture devoted to the methods of Thue-Siegel-Roth and Gelfond-Baker. Main ideas of proofs are given.

Taussky, Olga, *Sums of squares*, Amer. Math. Monthly, 77 (1970) 805-830.

> A wide-ranging article discussing sums of squares in many different mathematical contexts ranging from the definition of norm for quaternions and the theorem that every positive integer is the sum of four squares to a Galois-theory proof of the characterization of Pythagorean triangles. Winner of the Lester R. Ford award for expository writing.

Vandiver, H.S., *Fermat's last theorem--its history and the nature of the known results concerning it*, <u>Amer. Math. Monthly</u>, 53 (1946) 555-578.

Extensive survey requiring background in number theory.

3.3.3 RESEARCH

Bateman, Paul T. and Diamond, Harold G., *Asymptotic distribution of Beurling's generalized prime numbers*, in W.J. LeVeque, <u>Studies in Number Theory</u>, M.A.A., 1969, pp. 152-210.

Cassels, J.W.S., *Diophantine equations with special reference to elliptic curves*, <u>J. London Math. Soc.</u>, 41 (1966) 193-291.

"The object of this survey is to give a lowbrow account...stressing the relevance to concrete diophantine equations... ." This quotation must be interpreted in the context of a sophisticated subject.

Davis, Martin, *Hilbert's tenth problem is unsolvable*, <u>Amer. Math. Monthly</u>, 80 (1973) 233-269.

An exposition of the recent proof by Matiyasevič that there is no algorithm "which will tell of a given polynomial Diophantine equation with integer coefficients whether or not it has a solution in integers." Winner of the LeRoy P. Steele prize for distinguished exposition of outstanding research, and the Lester R. Ford award for expository writing.

Iyanaga, Shokichi, *Algebraic theory of numbers*, in A.H. Livermore, <u>Science in Japan</u>, AAAS, 1965, pp. 81-113.

Kaplansky, Irving, et al., *Number theory*, <u>Encyclopaedia Britannica</u>, 15th ed., 1974, Macropaedia V. 13, pp. 358-381.

An introductory section at the undergraduate level is followed by sections on algebraic number theory, analytic number theory, geometric and probabilistic number theory. These sections provide thorough accounts of the history and methodology of the various aspects of number theory, as well as a review of important results.

Lewis, D.J., *Diophantine equations: p-adic methods*, in W.J. LeVeque, <u>Studies in Number Theory</u>, M.A.A., 1969, pp. 25-75.

"...[A] detailed analysis of the relationship between p-adic and rational solvability of linear and quadratic equations... The remainder of the article is devoted to [an] exposition of Skolem's method... ."

Rademacher, Hans, *Trends in research: the analytic number theory*, Bull. Amer. Math. Soc., 48 (1942) 379-401.

   Comprehensive survey, long bibliography. High powered.

Robinson, Julia, *Diophantine decision problems*, in W.J. LeVeque, Studies in Number Theory, M.A.A., 1969, pp. 76-116.

   Shows that there exist exponential diophantine equations such that there is no method for deciding their solvability.

Wyman, B.F., *What is a reciprocity law?*, Amer. Math. Monthly, 79 (1972) 571-586; Correction, 80 (1973) 281.

   This paper blends reciprocity laws (quadratic, cyclotomic, etc.) with class field theory.

RELATED REFERENCES

Almgren, F.J., Jr. and Montgomery, H., *The 1974 Fields Medals (II): An analyst and number theorist*, Science, 186 (1974) 130-131. [1.3.3]

Ayoub, Raymond, *Euler and the zeta function*, Amer. Math. Monthly, 81 (1974) 1067-1086; Addendum, 82 (1975) 737. [1.2.2]

Bell, Eric Temple, *Gauss and the early development of algebraic numbers*, National Math. Magazine, 18 (1944) 188-204, 219-233. [1.2.2]

DeLong, Howard, *Unsolved problems in arithmetic*, Scientific American, 224 (March 1971) 50-60, 124. [2.5.1]

Hanson, Norwood Russell, *Number theory and physical theory: an analogy*, in R.S. Cohen and M.W. Wartofsky, Boston Studies in the Philosophy of Science, V. 2, Humanities Pr., 1965, pp. 93-119. [7.1.1]

Honsberger, Ross, Ingenuity in Mathematics, New Math. Libr., No. 23, Random House, 1970; M.A.A., 1975. [3.1.1]

Honsberger, Ross, Mathematical Gems, M.A.A., 1973. [3.1.1]

Kuratowski, Kazimierz, *Wacław Sierpiński (1882-1969)*, Acta Arith., 21 (1972) 1-5. [1.3.0]

Levinson, Norman, *Coding theory: a counter-example to G.H. Hardy's conception of applied mathematics*, <u>Amer. Math. Monthly</u>, 77 (1970) 249-258. [7.5.2]

Lieber, Lillian R., <u>Take a Number</u>, Ronald Pr., 1946. [1.7.0]

Mahoney, Michael S., *Fermat's mathematics: proofs and conjectures*, <u>Science</u>, 178 (1972) 30-36. [1.2.1]

Rademacher, Hans and Toeplitz, Otto, <u>The Enjoyment of Mathematics: Selections from Mathematics for the Amateur</u>, Princeton U. Pr., 1957. [1.1.1]

Stein, Sherman K., *Algebraic tiling*, <u>Amer. Math. Monthly</u>, 81 (1974) 445-462. [3.5.2]

3.4 Linear Algebra

3.4.1 ELEMENTARY

Beaumont, Ross A., *Determinants*, McGraw-Hill Encyclopedia of Science and Technology, 1960, V. 4, pp. 81-84.

Beaumont, Ross A., *Linear systems of equations*, McGraw-Hill Encyclopedia of Science and Technology, 1960, V. 7, pp. 523-525.

Beaumont, Ross A., *Matrix theory*, McGraw-Hill Encyclopedia of Science and Technology, 1960, V. 8, pp. 182-184.

Brauer, Alfred, *Matrix*, Encyclopedia Americana, 1976, V. 18, pp. 437-438.

Faddeev, D.K., *Linear algebra*, in A.D. Aleksandrov, et al., Mathematics--Its Content, Methods and Meaning, V. 3, M.I.T., 1963, pp. 37-96.
  Concise survey of elementary linear algebra.

Grinstein, Louise, *Determinant*, Encyclopedia Americana, 1976, V. 9, pp. 22-23.

Langer, R.E., *What are Eigen-werte?*, Amer. Math. Monthly, 50 (1943) 279-287.
  Examples from vibrating string, quantum mechanics, differential equations, Hilbert space, integral equations, and quadratic forms.

McShane, E.J., *Vector spaces and their applications*, in The Mathematical Sciences--A Collection of Essays for COSRIMS, M.I.T., 1969, pp. 84-96.
  The theme of this article is the use of the principal axis theorem in the applications of vector spaces to mechanics (with a brief nod to quantum mechanics), and the resultant identification of the resonant, or natural, frequencies of a body. As a simple guide the author first considers a straight steel rod of irregular cross section clamped at one end. The energy of displacement of the free end is then a quadratic function of two variables--the principal axis theorem selects coordinates in which there is no mixed term. Several pages are perforce devoted to elementary notions of vector spaces.

Murnaghan, F.D., *An elementary presentation of the theory of quaternions*, Scripta Math., 10 (1944) 37-49.
  Applications to spinning electrons, relativity, wave mechanics. Requires only elementary linear algebra and calculus.

Walsh, Bertram, *The scarcity of cross products on Euclidean spaces*, Amer. Math. Monthly, 74 (1967) 188-194.

If reasonable demands are made of "cross products", then only on R, R^3, and R^7 can any exist.

Wilansky, Albert, *Spectral decomposition of matrices for high school students*, Math. Magazine, 41 (1968) 51-59.

Winner of the Lester R. Ford award for expository writing.

3.4.2 ADVANCED

Halmos, Paul R., *Finite-dimensional Hilbert spaces*, Amer. Math. Monthly, 77 (1970) 457-464.

The author's theme is that there are many interesting questions in finite-dimensional linear algebra which could have arisen long ago but for which the impetus actually came recently from operator theory of infinite-dimensional Hilbert space. He discusses three: an algebraic characterization of pairs of subspaces of a finite-dimensional Hilbert space, a geometric characterization of linear transformations in terms of rotations and projections (dilatation theory), and lattices of invariant subspaces. Winner of the Lester R. Ford award for expository writing.

Linear and multilinear algebra, Encyclopaedia Britannica 15th ed., 1974, Macropaedia V. 1, pp. 507-518.

Paige, Lowell J., *Jordan algebras*, in A.A. Albert, Studies in Modern Algebra, M.A.A., 1963, pp. 144-186.

3.4.3 RESEARCH

Flanders, Harley, *Methods of proof in linear algebra*, Amer. Math. Monthly, 63 (1956) 1-15.

Kaplansky, Irving, *Lie algebras*, in T.L. Saaty, Lectures on Modern Mathematics, V. 1, Wiley, 1963, pp. 115-132.

Aimed at the mathematician, this article could serve well as an introduction to any standard monograph on the subject.

RELATED REFERENCES

Aris, Rutherford, *Chemical kinetics and the ecology of mathematics*, Amer. Scientist, 58 (1970) 419-428. [7.1.1]

Brennan, Jean F., *Exploiting sparse matrices*, [6.4.0]
IBM Research Reports, 6:1 (1970) 1-8.

Carter F.L., *Perspective drawing by numbers*, [7.6.1]
Math. Gazette, 53 (1969) 133-139.

Evans, Trevor, *Nonassociative number theory*, [3.3.1]
Amer. Math. Monthly, 64 (1957) 299-309.

Gale, David, *How to solve linear inequalities*, [6.3.1]
Amer. Math. Monthly, 76 (1969) 589-599.

Hawkins, Thomas, *Cauchy and the spectral* [1.2.2]
theory of matrices, Historia Math., 2 (1975)
1-29.

Hawkins, Thomas, *The theory of matrices in* [1.2.2]
the 19th century, in Proc. Inter. Cong. Math.
(1974), V. 2, Canad Cong. Math., 1975, pp. 561-
570.

Hoffman, Alan J., *Eigenvalues of graphs*, in [3.2.3]
D.R. Fulkerson, Studies in Graph Theory,
Part II, M.A.A., 1975, pp. 225-245.

Householder, Alston S., *Numerical analysis*, [6.4.2]
in T.L. Saaty, Lectures on Modern Mathematics,
V. 1, Wiley, 1963, pp. 59-97.

Isard, Walter and Kaniss, Phyllis, *The 1973* [7.4.0]
Nobel prize for economic science, Science,
182 (1973) 568-569, 571.

Lanczos, Cornelius, *Linear systems in self-* [4.5.2]
adjoint form, Amer. Math. Monthly, 65 (1958)
665-679.

Leslie, P.H., *On the use of matrices in cer-* [7.2.2]
tain population mathematics, Biometrika, 33
(1945) 183-212.

Parlett, Beresford, *Matrix eigenvalue prob-* [6.4.1]
lems, Amer. Math. Monthly, 72 (1965) Suppl.
pp. 59-66.

Scherk, Peter, *Some concepts of conformal* [5.1.2]
geometry, Amer. Math. Monthly, 67 (1960) 1-30.

Tewarson, R.P., *Computations with sparse* [6.4.2]
matrices, SIAM Review, 12 (1970) 527-543.

Varga, Richard S., *Iterative methods for solv-* [6.4.2]
ing matrix equations, Amer. Math. Monthly, 72
(1965) Suppl. pp. 67-74.

Weyl, Hermann, *Ramifications, old and new, of*   [4.5.2]
the eigenvalue problem, Bull. Amer. Math. Soc.,
56 (1950) 115-139.

Wilkinson, J.H., *Modern error analysis*, SIAM   [6.4.2]
Review, 13 (1971) 548-568.

Wilson, R.J., *An introduction to matroid*   [3.2.2]
theory, Amer. Math. Monthly, 80 (1973) 500-525.

3.5 GROUP THEORY

3.5.0 GENERAL

Artin, Emil, *The theory of braids*, Amer. Scientist, 38 (1950) 112-119; also in Math. Teacher, 52 (1959) 328-333.

Eddington, Sir Arthur Stanley, *The theory of groups*, in J.R. Newman, The World of Mathematics, V. 3, Simon and Schuster, 1956, pp. 1558-1573.

   A fair introduction to the generality and applicability of the group concept in mathematics. Written for laymen.

Grossman, Israel and Magnus, Wilhelm, Groups and Their Graphs, New Math. Libr., No. 14, Random House, 1964; M.A.A., 1975.

   Designed to present notions of group theory at a very elementary level. The role of the generators is portrayed here through the use of the Cayley diagram of a group, a directed graph whose vertices correspond to the elements of the group and whose (positive) edges correspond to multiplication by a generator.

Hammond, Allen L., *Sporadic groups: exceptions, or part of a pattern?*, Science, 181 (1973) 146, 148.

   Discovery by M. O'Nan of a sporadic simple group, with 460,815,505,920 elements, and the connection between sporadic groups and error-correcting codes.

Kestelman, H., *Wallpaper patterns*, in N.J. Hardiman, Exploring University Mathematics, V. 2, Pergamon, 1968, pp. 60-85.

   Excellent brief discussion of the five simplest patterns.

Mackey, George W., *Group theory and its significance for mathematics and physics*, Proc. Amer. Phil. Soc., 117 (1973) 374-380.

3.5.1 ELEMENTARY

Chandler, Bruce, *Group*, Encyclopedia Americana, 1976, V. 13, pp. 515-516.

Lieber, Lillian R., Galois and the Theory of Groups: A Bright Star in Mathesis, Galois Institute, 1932; 1956.

Shubnikov, A.V. and Koptsik, V.A., Symmetry in Science and Art, Plenum Pr., 1974.

Wightman, A.S., *Group theory*, McGraw-Hill Encyclopedia of Science and Technology, 1960, V. 6, pp. 282-285.

3.5.2 ADVANCED

Andrus, Jan F. and Butson, Alton T., *Ordered groups*, Amer. Math. Monthly, 70 (1963) 619-628.

Baylis, E.R., *Knots--a practical application of group theory*, Math. Gazette, 57 (1973) 311-320.

Elder, Barbara, *Paths and knots as geometric groups*, Pentagon, 28 (1968) 3-15.

Hall, Marshall, Jr., *Generators and relations in groups-- the Burnside problem*, in T.L. Saaty, Lectures on Modern Mathematics, V. 2, Wiley, 1964, pp. 42-92.

> When the original (1902) version of Burnside's problem in group theory appeared to be too intractable (it has since been solved in the negative) it was replaced by: if a group G is finitely generated and if every element of G has order dividing a fixed number n, is G necessarily finite? The answer is known to be "yes" when $n = 2, 3, 4$, or 6, and "no" for all odd $n \geq 4381$, the latter being a result of Novikov's which had only been announced when this paper was written. The paper surveys the "pre-Novikov" situation for the Burnside and restricted Burnside problems.

Hall, P., *Some word problems*, J. London Math. Soc., 33 (1958) 482-496.

> Survey of various such problems in algebra. No bibliography.

Mal'cev, A.I., *Groups and other algebraic systems*, in A.D. Aleksandrov, et al., Mathematics--Its Content, Methods and Meaning, V. 3, M.I.T., 1963, pp. 263-351.

> The first half of this article develops elementary group theory with emphasis on its relation to symmetry, e.g., the use of crystallographic groups to analyze symmetries of 3-space. There follow brief descriptions of a variety of other algebraic systems, some more or less limited to definitions: Lie groups and Lie algebras, topological groups, representations and characters of groups, quaternions, rings, ideals, and lattices.

Mayer, Arthur, *Rotations and their algebra*, SIAM Review, 2 (1960) 77-122.

> Includes differential calculus of rotations, resolver systems in analog computers, equations of motion of rigid bodies--but main attraction is good treatment of rotations, Euler angles, and relevant groups.

Mech, William P., *Graphs of groups*, <u>J. Undergraduate Math.</u>, 1 (1969) 97-110; 2 (1970) 37-49.

Passman, D.S., *What is a group ring?*, <u>Amer. Math. Monthly</u>, 83 (1976) 173-185.

"[For] finite groups, the group ring K[G] has historically been used as a tool of group theory. This is of course what the ordinary and modular character theory is all about... [If] G is infinite then neither the group theory nor the ring theory is particularly advanced and what becomes interesting here is the interplay between the two." The article discusses this latter case.

Stauduhar, Richard P., *The determination of Galois groups*, <u>Math. of Computation</u>, 27 (1973) 981-996.

A practical and relatively simple procedure for developing computer programs for the determination of the Galois groups of polynomials of degree less than or equal to 7. Gives two sample calculations.

Stein, Sherman K., *Algebraic tiling*, <u>Amer. Math. Monthly</u>, 81 (1974) 445-462.

This paper describes a variety of problems from number theory, the tiling of E^n, and coding theory which can be reformulated as problems concerning abelian groups. The principal example is a conjecture of Minkowski in number theory which was reformulated in terms of vectors and matrices, then in terms of tiling, and finally solved by Hajos in 1942 after he had reformulated it as a problem in abelian groups--a series of transitions of the problem "almost as startling as the metamorphosis of a caterpillar to a butterfly." Winner of the Lester R. Ford award for expository writing.

3.5.3 RESEARCH

Boone, William W., *The word problem*, <u>Annals of Math.</u>, 70 (1959) 207-265.

The word problem for groups is shown unsolvable.

Brauer, Richard, *Representations of finite groups*, in T.L. Saaty, <u>Lectures on Modern Mathematics</u>, V. 1, Wiley, 1963, pp. 133-175.

Cannon, John J., *Computers in group theory: a survey*, <u>Comm. Assoc. Comp. Mach.</u>, 12 (1969) 3-12.

Coset enumeration, subgroup lattices, automorphism groups, character tables, commutator calculus.

[3.5]

Carter, R.W., *Simple groups and simple Lie algebras*, <u>J. London Math. Soc.</u>, 40 (1965) 193-240.

>Survey of structure theory, details of classical examples, and succinct introduction to Lie algebras.

Chunikhin, S.A., *Some trends in the development of the theory of finite groups in recent years*, <u>Russian Math. Surveys</u>, 16:4 (1961) 29-46.

Curtis, Charles W., *The classical groups as a source of algebraic problems*, <u>Amer. Math. Monthly</u>, 74 (1967) Suppl. pp. 80-91.

>The author's aim is "...to describe the influence on the development of algebra of one of the great sources of algebraic problems and ideas--the classical linear groups." After mentioning the pioneering work of Dickson and the reworking and extensions of this by Dieudonné and Artin, Curtis discusses classification attempts for non-Abelian simple groups and the relevance to this classification of the classification of simple Lie algebras and simple Lie groups. Other topics mentioned are representation theory, algebraic groups, and finite groups of Lie type.

Feit, Walter, *The current situation in the theory of finite simple groups*, in <u>Actes Cong. Inter. Math.</u>, (1970) V. 1, Gauthier-Villars, 1971, pp. 55-93.

Fisher, Irwin and Struik, Ruth R., *Nil algebras and periodic groups*, <u>Amer. Math. Monthly</u>, 75 (1968) 611-623.

>In 1902 William Burnside posed his now-famous problem in group theory: if a group G is finitely generated and if every element of G has finite order, is G necessarily finite? This article contains the solution, by Golod, of a somewhat analogous problem in algebras (posed by Kurosh), and the corollary that the answer to the Burnside problem is no. The proof has the advantage of great simplicity vis-a-vis the (330 page) solution by Novikov of a stricter version of the Burnside problem, but even this expository account requires a fair background in algebra.

RELATED REFERENCES

Crowe, Donald W., *The geometry of African art [7.6.1]
I. Bakuba art*, <u>J. of Geometry</u>, 1 (1971) 169-182; *II. A catalog of Benin patterns*, <u>Historia Math.</u>, 2 (1975) 253-271.

Goldstein, Marie, *The historical development [1.2.2]
of group theoretical ideas in connection with Euclid's axiom of congruence*, <u>Notre Dame J. of Formal Logic</u>, 13 (1972) 331-349.

Group Theory   83   [3.5]

Hawkins, Thomas, *Hypercomplex numbers, Lie*   [1.2.3]
groups and the creation of group representa-
tion theory, Arch. Hist. Exact Sci., 8 (1972)
243-287.

Hawkins, Thomas, *New light on Frobenius'*   [1.2.3]
creation of the theory of group characters,
Arch. Hist. Exact Sci., 12 (1974) 217-243.

Hawkins, Thomas, *The origins of the theory*   [1.2.3]
of group characters, Arch. Hist. Exact Sci.,
7 (1971) 142-170.

Hilton, Peter J., *Localization in topology*,   [5.7.3]
Amer. Math. Monthly, 82 (1975) 113-131.

Infeld, Leopold, Whom the Gods Love,   [1.3.0]
Whittlesey House, 1948.

Kiernan, B. Melvin, *The development of Galois*   [1.2.1]
theory from Lagrange to Artin, Arch. Hist.
Exact Sci., 8 (1971) 40-154.

Ore, Oystein, Niels Henrik Abel, Mathemati-   [1.3.0]
cian Extraordinary, U. Minn. Pr., 1957;
Chelsea, 1974.

Weyl, Hermann, Symmetry, Princeton U. Pr.,   [1.4.1]
1952; excerpted in J.R. Newman, The World
of Mathematics, V. 1, Simon and Schuster,
1956, pp. 671-724.

Wigner, Eugene P., *Symmetry principles in old*   [7.1.2]
and new physics, Bull. Amer. Math. Soc., 74
(1968) 793-815.

3.6 ALGEBRAIC STRUCTURES

3.6.0 GENERAL

Boyer, Lee E., *Arithmetic*, Encyclopedia Americana, 1976, V. 2, pp. 293-297.

Cooley, H.R., *Algebra*, McGraw-Hill Encyclopedia of Science and Technology, 1960, V. 1, pp. 238-241.

Davis, Philip J., *Number*, Scientific American, 211 (September 1964) 50-59, 269; also in M. Kline, Mathematics in the Modern World, Freeman, 1968, pp. 89-97, 396.
> Somewhat difficult survey, including quaternions, matrices and transfinite numbers.

MacDuffee, C.C., Smith, David E. and LeVeque, William J., *Arithmetic*, Encyclopaedia Britannica, 15th ed., 1974, Macropaedia V. 1, pp. 1171-1178.

3.6.1 ELEMENTARY

Arnold, B.H., *Boolean algebra*, Encyclopedia Americana, 1976, V. 4, pp. 254-255.

Beaumont, Ross A., *Theory of equations*, McGraw-Hill Encyclopedia of Science and Technology, 1960, V. 5, pp. 46-47.

Birkhoff, Garrett, *Boolean algebra*, McGraw-Hill Encyclopedia of Science and Technology, 1960, V. 2, pp. 288-290.

Chandler, Bruce, *Field*, Encyclopedia Americana, 1976, V. 11, pp. 163-164.

Chen, Wai-Kai, *Boolean matrices and switching nets*, Math. Magazine, 39 (1966) 1-8.
> Winner of the Lester R. Ford award for expository writing.

Delone, B.N., *Algebra: theory of algebraic equations*, in A.D. Aleksandrov, et al., Mathematics--Its Content, Methods and Meaning, V. 1, M.I.T., 1963, pp. 261-310.
> Historical perspective, with a discussion of Galois groups and methods of approximating roots.

Elementary and multivariate algebra, Encyclopaedia Britannica, 15th ed., 1974, Macropaedia V. 1, pp. 499-507.
> The usual properties of algebraic operations are reviewed and illustrated geometrically. The remainder of the article is devoted

to a discussion of polynomials and rational functions, and the problem of solutions of systems of equations.

Hausner, Melvin, *Equation*, <u>Encyclopedia Americana</u>, 1976, V. 10, pp. 527-530.

Marcus, Marvin and Minc, Henryk, *Permanents*, <u>Amer. Math. Monthly</u>, 72 (1965) 577-591.
  Winner of the Lester R. Ford award for expository writing.

Rees, Mina and Shenton, Walter F., *Algebra*, <u>Encyclopedia Americana</u>, 1976, V. 1, pp. 555-562.

Richards, Ian, *Impossibility*, <u>Math. Magazine</u>, 48 (1975) 249-262.
  Seven mathematical puzzles and problems in which impossibility plays a role, including the fifteen-puzzle, the magic five-pointed star, the trisection of an angle, and the general quintic.

3.6.2 ADVANCED

Abian, A., *The Stone space of a Boolean ring*, <u>L'Enseignement Math.</u>, 11 (1965) 194-198,
  Expository paper on Stone's 1936-37 papers.

Arbib, Michael A. and Manes, Ernest G., *Machines in a category: an expository introduction*, <u>SIAM Review</u>, 16 (1974) 163-192.
  Category theory applied to automata theory.

Bruck, R.H., *What is a loop?*, in A.A. Albert, <u>Studies in Modern Algebra</u>, M.A.A., 1963, pp. 59-99.

Cohn, P.M., *Rings of fractions*, <u>Amer. Math. Monthly</u>, 78 (1971) 596-615.
  In 1930 Ore showed how to embed certain non-commutative rings in (non-commutative) fields, thus generalizing the corresponding standard result for integral domains. This paper reviews Ore's work and brings the subject up to date. Winner of the Lester R. Ford award for expository writing.

Cohn, P.M., *Unique factorization domains*, <u>Amer. Math. Monthly</u>, 80 (1973) 1-18; Correction, 80 (1973) 1115.
  The author develops the properties of the class of rings called domains as an example of how ring-theoretical properties can be illustrated by geometrical ideas.

Finkbeiner, Daniel T., *Vector and tensor analysis*, <u>Encyclopaedia Britannica</u>, 15th ed., 1974, Macropaedia V. 1, pp. 791-799.

> A review of vector algebra is followed by a discussion of algebras in general, and algebras of tensors (including the associated Grassman and Clifford algebras) in particular.

Hilton, Peter J., *Categories and functors*, <u>Math. Teaching</u>, 50 (1970) 30-34.

> Elementary introduction, suitable for an abstract algebra class.

Kurosh, A.G., Livshits, A. Kh. and Shul'Geifer, E.G., *Foundations of the theory of categories*, <u>Russian Math. Surveys</u>, 15:6 (1960) 1-46.

MacLane, Saunders, *Some recent advances in algebra*, <u>Amer. Math. Monthly</u>, 46 (1939) 3-19; also in A.A. Albert, <u>Studies in Modern Algebra</u>, M.A.A., 1963, pp. 9-34.

> Winner of the Chauvenet Prize for expository writing in mathematics.

Yale, Paul B., *Automorphisms of the complex numbers*, <u>Math. Magazine</u>, 39 (1966) 135-141.

> Winner of the Lester R. Ford award for expository writing.

Zassenhaus, Hans J., *On the fundamental theorem of algebra*, <u>Amer. Math. Monthly</u>, 74 (1967) 485-497.

> Winner of the Lester R. Ford award for expository writing.

3.6.3 RESEARCH

Birkhoff, Garrett, et al., *Algebraic structures*, <u>Encyclopaedia Britannica</u>, 15th ed., 1974, Macropaedia V. 1, pp. 518-558.

> Algebraic structures are organized here as lattice theory, groups, fields, rings, categories, homological algebra, and universal algebra--along with numerous subclassifications. The level ranges from elementary to Serre's use of Leray's spectral sequences to relate the homology groups and homotopy groups of a topological space. Most of the material is, of course, more explicitly algebraic and outlines much of what would be covered in an undergraduate algebra course plus at least part of a graduate course. Group representations are treated separately (<u>ibid</u>., Macropaedia V. 1, pp. 752 - 757) under Fourier Analysis.

Bliss, Gilbert A., *Algebraic functions and their divisors*, Annals of Math., 26 (1924-25) 95-124.

> Winner of the Chauvenet Prize for expository writing in mathematics.

Carrell, James B. and Dieudonné, Jean A., *Invariant theory, old and new*, in Advances in Math., 4 (1969) 1-80.

> Winner of the LeRoy P. Steele prize for distinguished exposition of outstanding research.

Cohn, P.M., *Algebra and language theory*, Bull. London Math. Soc., 7 (1975) 1-29.

> "Since (language theory) is mathematical at heart, and perhaps not as widely known among mathematicians as it deserves to be, it seemed worthwhile to present a brief survey of the main results." By presupposing a mathematical background, the author is able to present a briefer account of the subject than is customary, and to emphasize the mathematical content.

Cohn, P.M., *Free associative algebras*, Bull. London Math. Soc., 1 (1969) 1-39.

> "Free commutative and associative algebras--better known as polynomial rings--play an important role in commutative ring theory and in algebraic geometry. By contrast, free associative algebras have up to now played a relatively minor role for noncommutative rings. This is partly due...to the difficulty of the questions raised... ."

Curtis, Charles W., *The four and eight square problem and division algebras*, in A.A. Albert, Studies in Modern Algebra, M.A.A., 1963, pp. 100-125.

> Properties of quaternions and Cayley numbers, with emphasis on the implications of their having norms. Includes relationships with several algebras (normed, division, alternative, Lie, Jordan) as well as relationships with various topological notions.

Glushkov, V.M., *The abstract theory of automata*, Russian Math. Surveys, 16:5 (1961) 1-54.

Goldie, A.W., *Some aspects of ring theory*, Bull. London Math. Soc., 1 (1969) 129-154.

> In the first (1930) edition of his algebra text, van der Waerden reported as still open the question of whether a non-commutative integral domain could be embedded in a division ring. In 1936 the question was settled in the negative by a counterexample due to Mal'cev. Positive results had been obtained by Wedderburn (the quotient method works for Euclidean domains) and Ore. Beginning with these facts, Goldie's paper surveys subsequent developments in some aspects of ring theory.

Lam, T.Y. and Siu, M.K., K_0 and K_1--an introduction to algebraic K-theory, Amer. Math. Monthly, 82 (1975) 329-364.

> This survey of K_0 and K_1 is at a more advanced level than most articles in the Monthly. Readers who don't have the algebraic background for it might still enjoy the thumbnail history of K-theory on pp. 362-3.

MacLane, Saunders, *Categorical algebra*, Bull. Amer. Math. Soc., 71 (1965) 40-106.

> A short course on category theory.

MacLane, Saunders, *Modular fields*, Amer. Math. Monthly, 47 (1940) 259-274.

> Winner of the Chauvenet Prize for expository writing in mathematics.

MacLane, Saunders, *Some additional advances in algebra*, in A.A. Albert, Studies in Modern Algebra, M.A.A., 1963, pp. 35-58.

> A sequel to the author's 1939 article *Some recent advances in algebra* (ibid., pp. 9-34) which begins with a very readable account of recent developments in finite group theory and goes on to such topics as local rings, modules and tensor products, homological algebra, and category theory.

Samuel, Pierre, *Unique factorization*, Amer. Math. Monthly, 75 (1968) 945-952.

> A condensed survey of unique factorization domains with necessary side comments on principal ideal rings, Euclidean domains, etc., en route to more advanced topics. The non-specialist will want to have a modern algebra text at hand to illuminate the article, but may find that the article also illuminates the text. Winner of the Lester R. Ford award for expository writing.

RELATED REFERENCES

Birkhoff, Garrett, *Current trends in algebra*, Amer. Math. Monthly, 80 (1973) 760-782; Correction, 81 (1974) 746.   [1.2.2]

Bott, Raoul, *The periodicity theorem for the classical groups and some of its applications*, Advances in Math., 4 (1970) 353-411.   [5.7.3]

Boyd, John Paul, *The algebra of group kinship*, J. Math. Psych., 6 (1969) 139-167.   [7.3.2]

Brun, Viggo, *Euclidean algorithms and musical theory*, L'Enseignement Math., 10 (1964) 125-137. [7.6.0]

Collins, G.E., *Computer algebra of polynomials and rational functions*, Amer. Math. Monthly, 80 (1973) 725-755. [6.5.2]

Cowan, Thaddeus M., *The theory of braids and the analysis of impossible figures*, J. Math. Psych., 11 (1974) 190-212. [7.6.2]

Eilenberg, Samuel, *The algebraization of mathematics*, in The Mathematical Sciences--A Collection of Essays for COSRIMS, M.I.T., 1969, pp. 153-160. [1.5.0]

Fisher, Charles S., *The death of a mathematical theory: a study in the sociology of knowledge*, Arch. Hist. Exact Sci., 3 (1966) 137-159. [1.2.1]

Halmos, Paul R., *The basic concepts of algebraic logic*, Amer. Math. Monthly, 63 (1956) 363-387. [2.6.3]

Hille, Einar, *What is a semi-group?*, in I.I. Hirschman, Jr., Studies in Real and Complex Analysis, M.A.A., 1965, pp. 55-66. [4.6.3]

Koppelman, Elaine, *The calculus of operations and the rise of abstract algebra*, Arch. Hist. Exact Sci., 8 (1971) 155-242. [1.2.1]

Levinson, Norman, *Coding theory: a counterexample to G.H. Hardy's conception of applied mathematics*, Amer. Math. Monthly, 77 (1970) 249-258. [7.5.2]

Lifshits, V.N. and Sadovskii, L.E., *Algebraic models of computing machines*, Russian Math. Surveys, 27:3 (1972) 87-135. [6.5.3]

Monk, J. Donald, *Connections between combinatorial theory and algebraic logic*, in A. Daigneault, Studies in Algebraic Logic, M.A.A., 1974, pp. 58-91. [2.6.3]

Nový, Luboš, The Origins of Modern Algebra, Noordhoff, 1974. [1.2.2]

Passman, D.S., *What is a group ring?*, Amer. Math. Monthly, 83 (1976) 173-185. [3.5.2]

Robinson, Abraham, *Model theory as a frame-work for algebra*, in M.D. Morley, Studies in Model Theory, M.A.A., 1973, pp. 134-157. [2.6.2]

Seebach, J. Arthur, Jr., Seebach, Linda A. and Steen, Lynn Arthur, *What is a sheaf?*, Amer. Math. Monthly, 77 (1970) 681-703. [5.6.2]

4.1 ELEMENTARY CALCULUS

4.1.1 ELEMENTARY

Boas, Ralph P., Jr., *Inequalities for the derivatives of polynomials*, Math. Magazine, 42 (1969) 165-174.
  Winner of the Lester R. Ford award for expository writing.

Butchart, J.H. and Moser, Leo, *No calculus, please*, Scripta Math., 18 (1952) 221-236.
  Solving calculus problems without calculus.

Clark, Colin W., *Some socially relevant applications of elementary calculus*, Two-Year College Mathematics Journal, 4 (1973) 1-15.

Coolidge, J.L., *The number e*, Amer. Math. Monthly, 57 (1950) 591-602; also in T.M. Apostol, Selected Papers on Calculus, M.A.A., 1969, pp. 8-19.

Eberlein, W.F., *The elementary transcendental functions*, Amer. Math. Monthly, 61 (1954) 386-392; also in T.M. Apostol, Selected Papers on Calculus, M.A.A., 1969, pp. 126-133.
  Explicitly parallel treatment of circular and hyperbolic functions, first from a geometric and then from an analytic perspective. Excellent calculus supplement.

Feller, William, *A direct proof of Stirling's formula*, Amer. Math. Monthly, 74 (1967) 1223-1225; Correction, 75 (1968) 518.
  An elementary proof of Stirling's approximation to the factorial function.

Fletcher, T.J., *Doing without calculus*, Math. Gazette, 55 (1971) 4-17.
  Suggests ways in which some of the traditional max-min problems of calculus may be studied by other methods--even the "ladder round the corner" problem! The fundamental principle in each problem is the inequality between the arithmetic and geometric means.

Giblin, P.J., *What is an asymptote?*, Math. Gazette, 56 (1972) 274-284.
  Shows that there are several different possible definitions of an asymptote.

Graves, Lawrence M., *Integration*, McGraw-Hill Encyclopedia of Science and Technology, 1960, V. 7, pp. 166-174.

Kline, Morris, *Calculus*, Encyclopedia Americana, 1976, V. 5, pp. 163-177.

Lavrent'ev, M.A. and Nikol'skiĭ, S.M., *Analysis*, in A.D. Aleksandrov, et al., Mathematics--Its Content, Methods and Meaning, V. 1, M.I.T., 1963, pp. 65-180.

   A systematic survey of elementary calculus, ranging from limits and continuity to Taylor's series and functions of several variables.

Mancill, Julian D., *On the elementary transcendental functions*, in K.O. May, Lectures on Calculus, Holden-Day, 1967, pp. 15-45.

   A unified presentation of the elementary calculus of the exponential, logarithmic and trigonometric functions.

Richmond, D.E., *Areas and volumes without limit processes*, Amer. Math. Monthly, 73 (1966) 477-483.

   Areas and volumes under polynomial graphs.

Sawyer, W.W., What is Calculus About?, New Math. Libr., No. 2, Random House, 1961; M.A.A., 1975.

Taylor, Angus E., *Differential and integral calculus*, McGraw-Hill Encyclopedia of Science and Technology, 1960, V. 2, pp. 401-403.

Taylor, Angus E., *Differentiation*, McGraw-Hill Encyclopedia of Science and Technology, 1960, V. 4, pp. 128-132.

Thurston, Hugh, *What exactly is dy/dx?*, Educ. Studies Math., 4 (1972) 358-367.

   Pinpoints the logical deficiencies in the definition of dy/dx; resolves the difficulty in a discussion of one-dimensional manifolds in the plane.

4.1.2 ADVANCED

Rosenlicht, Maxwell, *Integration in finite terms*, Amer. Math. Monthly, 79 (1972) 963-972.

   "Can the indefinite integral of an explicitly given function of one variable always be expressed 'explicitly' (or 'in closed form', or 'in finite terms')?" This article contains the precise statement and proof of the negative answer to this question.

RELATED REFERENCES

Baron, Margaret E., *The Origins of the Infinitesimal Calculus*, Pergamon, 1969. [1.2.1]

Beckenbach, Edwin F. and Bellman, Richard, *An Introduction to Inequalities*, New Math. Libr., No. 3, Random House, 1961; M.A.A., 1975. [6.3.1]

Bell, Eric Temple, Newton after three centuries, *Amer. Math. Monthly*, 49 (1942) 553-575. [1.2.1]

Bos, H.J.M., Differentials, higher-order differentials and the derivative in the Leibnizian calculus, *Arch. Hist. Exact Sci.*, 14 (1974) 1-90. [1.2.1]

Desanti, Jean T., From Cauchy to Riemann, or the birth of the theory of real functions, in F. LeLionnais, *Great Currents of Mathematical Thought*, V. 1, Dover, 1971, pp. 181-190. [1.2.1]

Drake, Stillman, Galileo's discovery of the law of free fall, *Scientific American*, 228 (May 1973) 84-92, 120. [1.2.1]

Drake, Stillman, Mathematics and discovery in Galileo's physics, *Historia Math.*, 1 (1974) 129-150. [1.2.1]

Drake, Stillman and MacLachlan, James, Galileo's discovery of the parabolic trajectory, *Scientific American*, 232 (March 1975) 102-110, 132. [7.1.0]

Grabiner, Judith V., Is mathematical truth time-dependent?, *Amer. Math. Monthly*, 81 (1974) 354-365. [1.5.1]

Greenspan, H.P., Applied mathematics as a science, *Amer. Math. Monthly*, 68 (1961) 872-880. [1.4.1]

Hahn, Hans, The crisis in intuition, in J.R. Newman, *The World of Mathematics*, V. 3, Simon and Schuster, 1956, pp. 1956-1976. [1.5.0]

Robinson, Abraham, Some thoughts on the history of mathematics, *Compositio Math.*, 20 (1968) 188-193. [2.4.1]

Robinson, Abraham, *The metaphysics of the cal-*   [2.4.1]
culus, in J. Hintikka, The Philosophy of
Mathematics, Oxford U. Pr., 1969, pp. 153-
163; also in I. Lakatos, Problems in the
Philosophy of Mathematics, North-Holland,
1967, pp. 28-40.

Rosenthal, Arthur, *The history of calculus*,   [1.2.1]
Amer. Math. Monthly, 58 (1951) 75-86.

Thurston, Hugh, *Tangents: an elementary sur-*   [5.1.1]
vey, Math. Magazine, 42 (1969) 1-11.

Taylor, Angus E., *The differential: nineteenth*   [1.2.1]
and twentieth century developments, Arch. Hist.
Exact Sci., 12 (1974) 355-383.

Wilson, Curtis, *How did Kepler discover his*   [1.2.0]
first two laws?, Scientific American, 226
(March 1972) 92-106, 126.

Zippin, Leo, Uses of Infinity, New Math. Libr.,   [2.1.0]
No. 7, Random House, 1962; M.A.A., 1975.

4.2 Advanced Calculus

4.2.1 ELEMENTARY

Cunningham, Frederic, Jr., *Taking limits under the integral sign*, Math. Magazine, 40 (1967) 179-186.

   Winner of the Lester R. Ford award for expository writing.

Franklin, Philip, *Series*, McGraw-Hill Encyclopedia of Science and Technology, 1960, V. 12, pp. 189-196.

Herz, Carl S., *Fourier series and integrals*, McGraw-Hill Encyclopedia of Science and Technology, 1960, V. 5, pp. 487-490.

Lass, Harry, *Calculus of vectors*, McGraw-Hill Encyclopedia of Science and Technology, 1960, V. 2, pp. 410-414.

Protter, M.H., *Potentials*, McGraw-Hill Encyclopedia of Science and Technology, 1960, V. 10, pp. 537-539.

Radó, Tibor, *What is the area of a surface?*, Amer. Math. Monthly, 50 (1943) 139-141.

   Discusses integral formula and its limitations, with an example of Schwarz.

Ravetz, Jerome R., *Fourier series*, Encyclopedia Americana, 1976, V. 11, pp. 657-658.

Rhodes, F., *$1 - 1 + 1 - 1 + \ldots = 1/2$?*, Math. Gazette, 55 (1971) 298-305.

   A lead-in to Cesàro summability.

Sagan, Hans, *Area and integration*, in K.O. May, Lectures on Calculus, Holden-Day, 1967, pp. 61-72.

   Brief, lucid discussion, including some strange cases where area is hard to define.

Tahta, D.G., *A startling discovery*, Math. Teaching, 45 (1968) 33-37.

   Examples of startling curves, narrated in historical perspective. Development of the concept of continuity, Bolzano's continuous infinitely kinky (non-differentiable, nowhere smooth) curve, Abel's continuous curves whose limit is not continuous, Weierstrass's continuous nowhere differentiable curve.

Thurston, Hugh, *Series*, Encyclopedia Americana, 1976, V. 24, pp. 576-580.

4.2.2 ADVANCED

Boas, Ralph P., Jr., *Inversion of Fourier and Laplace transforms*, <u>Amer. Math. Monthly</u>, 69 (1962) 955-960.

>Crude heuristic derivations of basic formulas for Fourier transforms proceed by a limiting process from corresponding formulas for Fourier series. Boas presents an "elementary" but rigorous way of making this step. His method is an adaptation of a proof due to A. Weil in a more general setting.

Brand, Louis, *A division algebra for sequences and its associated operational calculus*, <u>Amer. Math. Monthly</u>, 71 (1964) 719-728.

>Winner of the Lester R. Ford award for expository writing.

Bruckner, Andrew M., *Derivatives: why they elude classification*, <u>Math. Magazine</u>, 49 (1976) 5-11.

Bruckner, Andrew M. and Leonard, J.L., *Derivatives*, <u>Amer. Math. Monthly</u>, 73 (1966) Suppl. pp. 24-56.

>A lengthy survey of theorems and unsolved problems concerning the derivative of a function of one real variable, including variations and generalizations. The authors have striven to keep it readable by excluding topics which would require much preliminary machinery. Includes an extensive bibliography.

Buck, R.C., *Topology and analysis*, <u>Math. Magazine</u>, 40 (1967) 71-74.

>Illustration of how topological concepts illuminate analysis and vice-versa, with emphasis on the topology of pointwise convergence versus the topology of uniform convergence. Begins with some examples from elementary calculus.

Cesari, Lamberto, *Recent results in surface area theory*, <u>Amer. Math. Monthly</u>, 66 (1959) 173-192.

Cesari, Lamberto, *Surface area*, in S.S. Chern, <u>Studies in Global Geometry and Analysis</u>, M.A.A., 1967, pp. 123-146.

Darst, R.B., *Some Cantor sets and Cantor functions*, <u>Math. Magazine</u>, 45 (1972) 2-7.

Feller, William, *On Müntz' theorem and completely monotone functions*, <u>Amer. Math. Monthly</u>. 75 (1968) 342-350.

>The Weierstrass approximation theorem asserts that every (real) continuous function on [0, 1] is the uniform limit of an appropriate sequence of polynomials. Suppose that we complicate the problem by ruling out selected powers of x, e.g., all odd powers. Can we still approximate uniformly with polynomials us-

ing the remaining powers of x? Assuming that we keep the zero'th power, so as to have constants, the answer is very neat: if and only if the sum of the reciprocals of the non-zero eligible powers is a divergent series. This is a special case of a 1914 theorem of Müntz. The author gives a simple proof of Müntz's theorem and describes its relationship with other results.

Hardy, G.H., *The integral $\int_0^\infty (\sin x/x)dx$*, <u>Math. Gazette</u>, 5 (1909); reprinted, ibid., 55 (1971) 152-158.

The immortal G.H. grades several other mathematicians on the difficulty involved in their evaluations of this integral, with bonus penalties added "in a more capricious way" for artificiality, complexity, etc. A most amusing and entertaining romp.

Hirschman, I.I., Jr. and Widder, D.V., *The Laplace transform, the Stieltjes transform, and their generalizations*, in I.I. Hirschman, Jr., <u>Studies in Real and Complex Analysis</u>, M.A.A., 1965, pp. 67-89.

The opening sections of this article give a clear description of the Laplace transform, with an indication of its uses in, e.g., the solution of differential equations. The article then moves on to the inversion formulas of the Laplace and Stieltjes transforms. (The Stieltjes transform can be described as an iterated Laplace transform.)

Johnson, Paul and Redheffer, Raymond, *Scrambled series*, <u>Amer. Math. Monthly</u>, 73 (1966) 822-828.

Invariance of sum under rearrangements and "scrambling" of summands. Also treats the notions of length, area, and volume.

Murray, J.D., *Approximate methods in mathematics*, <u>Math. Spectrum</u>, 6 (1973-74) 19-24.

Use of divergent series to evaluate integrals.

Niven, Ivan M., *Formal power series*, <u>Amer. Math. Monthly</u>, 76 (1969) 871-889.

"Our purpose is to develop a systematic theory of formal power series. Such a theory is known, or at least presumed, by many writers on mathematics, who use it to avoid questions of convergence in infinite series. What is done here is to formulate the theory on a proper logical basis and thus to reveal the absence of the convergence question. Thus 'hard' analysis can be replaced by 'soft' analysis in many applications. ...[M]any examples of the use of formal power series could be cited from the literature... ." Winner of the Lester R. Ford award for expository writing.

Shiu, P., *How slowly can a series converge?*, <u>Math. Gazette</u>, 56 (1972) 285-288.

West, Jerry L., *The Cantor set*, <u>Pi Mu Epsilon Journal</u>, 5 (1970) 119-123.

[4.2] 98 Analysis:

Wyler, Oswald, *Exterior differential calculus and Maxwell's equations*, in K.O. May, <u>Lectures on Calculus</u>, Holden-Day, 1967, pp. 147-165.

RELATED REFERENCES

Aleksandrov, A.D., *Curves and surfaces*, in A.D. Aleksandrov, et al., <u>Mathematics--Its Content, Methods and Meaning</u>, V. 2, M.I.T., 1963, pp. 57-117. [5.3.1]

Callahan, James, *Singularities and plane maps*, <u>Amer. Math. Monthly</u>, 81 (1974) 211-240. [5.6.2]

Christie, Dan E., *Some thermodynamic properties of a mapping*, <u>J. Franklin Inst.</u>, 267 (1959) 119-133. [7.1.2]

Coolidge, J.L., *The lengths of curves*, <u>Amer. Math. Monthly</u>, 60 (1953) 89-93. [1.2.1]

Davis, Philip J., *Leonard Euler's integral: a historical profile of the gamma function*, <u>Amer. Math. Monthly</u>, 66 (1959) 849-869. [1.2.2]

Flanders, Harley, *A proof of Minkowski's inequality for convex curves*, <u>Amer. Math. Monthly</u>, 75 (1968) 581-593. [6.3.2]

Flanders, Harley, *Differential forms*, in S.S. Chern, <u>Studies in Global Geometry and Analysis</u>, M.A.A., 1967, pp. 57-95. [5.3.2]

Glasser, M. Lawrence, *The summation of series*, <u>SIAM J. Math. Anal.</u>, 2 (1971) 595-600. [6.4.2]

Mayer, Arthur, *Rotations and their algebra*, <u>SIAM Review</u>, 2 (1960) 77-122. [3.5.2]

Nikol'skiĭ, S.M., *Approximations of functions*, in A.D. Aleksandrov, et al., <u>Mathematics--Its Content, Methods and Meaning</u>, V. 2, M.I.T., 1963, pp. 265-302. [6.4.1]

Schoenberg, I.J., *The elementary cases of Landau's problem of inequalities between derivatives*, <u>Amer. Math. Monthly</u>, 80 (1973) 121-158. [6.4.2]

Walsh, Bertram, *The scarcity of cross products on Euclidean spaces*, <u>Amer. Math. Monthly</u>, 74 (1967) 188-194. [3.4.1]

4.3 REAL ANALYSIS

4.3.1 ELEMENTARY

Blumenthal, Leonard M., "*A paradox, a paradox, a most ingenious paradox*," Amer. Math. Monthly, 47 (1940) 346-353.

　Following a gentle introduction, Blumenthal bombs the reader's intuition with the major paradoxes of measure and set theory--Sierpiński-Mazurkiewicz, Hausdorff, Banach-Tarski.

Kaufman, Hyman, *Real analysis*, Encyclopaedia Britannica, 15th ed., 1974, Macropaedia V. 1, pp. 772-791.

　Discussion of the origins and concepts of analysis, the development and properties of the real number system, functions (of one variable) and differential calculus, measure and integral calculus.

Stečkin, S.B., *Theory of functions of a real variable*, in A.D. Aleksandrov, et al., Mathematics--Its Content, Methods and Meaning, V. 3, M.I.T., 1963, pp. 3-36.

　An exposition of the elements of real analysis starting with the construction of the real numbers and various completeness criteria ("principles of continuity") and concluding with Lebesgue measure on the reals, the Lebesgue integral of a bounded measurable function on a finite interval, and some hints as to why the Lebesgue integral is useful in the study of trigonometric series.

Ulam, Stanislaw M., *What is measure?*, Amer. Math. Monthly, 50 (1943) 597-602.

　Historical exposition with a clear, concise introduction to Borel sets, Lebesgue measure and related problems.

4.3.2 ADVANCED

Botts, Truman, *Probability theory and the Lebesgue integral*, Math. Magazine, 42 (1969) 105-111.

　A review of probability motivates the notion of a sigma field of sets, exemplified by the Lebesgue measurable sets. This leads to a comparison of the Riemann and Lebesgue integrals and a mathematical example where the latter is a natural tool but the former is inadequate.

Doob, J.L., *Probability in function space*, Bull. Amer. Math. Soc., 53 (1947) 15-30.

> Self-contained introduction.

Erdélyi, Arthur, *From delta functions to distributions*, in E.F. Beckenbach, Modern Mathematics for the Engineer, 2nd Ser., McGraw-Hill, 1961, pp. 5-50.

> This article presents, side by side, developments of the Mikusiński operational calculus and of the theory of distributions--the two main methods of explicating the Dirac delta function. It could serve well as an introduction to the author's book on the Mikusiński calculus.

Goffman, Casper, *Completeness of the real numbers*, Math. Magazine, 47 (1974) 1-8.

> Relations among three different notions of completeness of the real numbers: in the sense of Dedekind (least upper bound property), in the sense of Cantor (Cauchy sequences converge), and as a totally ordered group.

Green, J.W., *Recent applications of convex functions*, Amer. Math. Monthly, 61 (1954) 449-454.

> "Recent" refers to the years 1946-1954.

Hewitt, Edwin, *The role of compactness in analysis*, Amer. Math. Monthly, 67 (1960) 499-516.

> Lucid exposition, centered on a few major results of classical analysis, of compactness as an extension of finiteness. Long bibliography.

Jackson, Dunham, *The convergence of Fourier series*, Amer. Math. Monthly, 41 (1934) 67-84.

> Winner of the Chauvenet Prize for expository writing in mathematics.

Lebesgue, Henri, *The development of the integral concept*, in H. Lebesgue, Measure and the Integral, Holden-Day, 1966, pp. 178-194.

> A clear introductory exposition by the creator himself. He discusses both the motivation and special characteristics which distinguish his definition from Riemann's.

McShane, E.J., *A theory of limits*, in R.C. Buck, Studies in Modern Analysis, M.A.A., 1962, pp. 7-29.

McShane, E.J., *A unified theory of integration*, Amer. Math. Monthly, 80 (1973) 349-359.

McShane, E.J., *Partial orderings and Moore-Smith limits*, Amer. Math. Monthly, 59 (1952) 1-11.

  Winner of the Chauvenet Prize for expository writing in mathematics.

Munroe, M.E., *Bringing calculus up to date*, Amer. Math. Monthly, 65 (1958) 81-90.

  Differential geometry and real function theory.

Nashed, M.Z., *Some remarks on variations and differentials*, Amer. Math. Monthly, 73 (1966) Suppl. pp. 63-76.

  Winner of the Lester R. Ford award for expository writing.

Newns, W.F., *Functional dependence*, Amer. Math. Monthly, 74 (1967) 911-920.

  Winner of the Lester R. Ford award for expository writing.

Rosenthal, Arthur, *What are set functions?*, Amer. Math. Monthly, 55 (1948) 14-20.

  Properties of set functions, and their use in integration and differentiation.

Schaefer, H.H., *A brief introduction to the Lebesgue-Stieltjes integral*, in I.I. Hirschman, Jr., Studies in Real and Complex Analysis, M.A.A., 1965, pp. 90-123.

Serrin, James, *On the area of curved surfaces*, Amer. Math. Monthly, 68 (1961) 435-440.

  "It is a common impression that if one is to say anything rigorously correct about the area of a surface he must first spend several months of his life with some of the most difficult books in the mathematical library. ...[No calculus texts] present more than a heuristic account. ...Starting from Lebesgue's definition of area, we shall show that the fundamental results of the theory (for surfaces of the form $z = f(x, y)$) follow at once from a single inequality whose geometric content is roughly that integral smoothing is an area shrinking operation. ...Lebesgue's happy idea was to modify the [definition via approximation polyhedra] by using the lower limit of the areas of approximating polyhedra."

Tolsted, Elmer, *An elementary derivation of the Cauchy, Hölder, and Minkowski inequalities from Young's inequality*, Math. Magazine, 37 (1964) 2-12.

  Winner of the Lester R. Ford award for expository writing.

Whyburn, Gordon T., *Topological analysis*, Bull. Amer. Math. Soc., 62 (1956) 204-218.

  Surveys many interactions between topology and analysis.

4.3.3 RESEARCH

Bruckner, Andrew M., *Differentiation of integrals*, <u>Amer. Math. Monthly</u>, 78 (1971) Suppl. pp. 1-51.

    An extensive exposition of "work done in the last thirty-five years on various questions concerning the differentiation of integrals in spaces more general than the Euclidean line R. Perhaps the unifying notion would be that of the Fundamental Theorem of Calculus in such spaces."

Goffman, Casper and Waterman, Daniel, *Some aspects of Fourier series*, <u>Amer. Math. Monthly</u>, 77 (1970) 119-133.

Hildebrandt, T.H., *Integration in abstract spaces*, <u>Bull. Amer. Math. Soc.</u>, 59 (1953) 111-139.

    Semihistorical survey of the various generalizations of the integral beginning with Lebesgue.

Hildebrandt, T.H., *The Borel theorem and its generalizations*, <u>Bull. Amer. Math. Soc.</u>, 32 (1926) 423-474.

    Winner of the Chauvenet Prize for expository writing in mathematics.

Jackson, Dunham, *Orthogonal trigonometric sums*, <u>Annals of Math.</u>, 34 (1933) 799-814.

    Winner of the Chauvenet Prize for expository writing in mathematics.

Jackson, Dunham, *Series of orthogonal polynomials*, <u>Annals of Math.</u>, 34 (1933) 527-545.

    Winner of the Chauvenet Prize for expository writing in mathematics.

Meisters, G.H. and Monk, J. Donald, *Construction of the reals via ultrapowers*, <u>Rocky Mountain J. Math.</u>, 3 (1973) 141-158.

    Construction of the reals from the rationals.

Phillips, Keith L., *The maximal theorems of Hardy and Littlewood*, <u>Amer. Math. Monthly</u>, 74 (1967) 648-660.

    Winner of the Lester R. Ford award for expository writing.

Segal, Irving, *Algebraic integration theory*, <u>Bull. Amer. Math. Soc.</u>, 71 (1965) 419-489.

RELATED REFERENCES

Aberth, Oliver, *Analysis in the computable number field*, <u>J. Assoc. Comp. Mach.</u>, 15 (1968) 275-299. [2.3.2]

Bochner, Salomon, *The rise of functions*, <u>Rice Univ. Studies</u>, 56:2 (1970) 3-21. [1.2.2]

Brush, Stephen G., *Foundations of statistical mechanics 1845-1915*, <u>Arch. Hist. Exact Sci.</u>, 4 (1967) 145-183. [7.1.2]

Coppel, W.A., *J.B. Fourier--On the occasion of his two hundredth birthday*, <u>Amer. Math. Monthly</u>, 76 (1969) 468-483. [1.2.3]

Grattan-Guinness, Ivor, <u>Joseph Fourier, 1768-1830</u>, M.I.T. Pr., 1972. [1.3.0]

Grattan-Guinness, Ivor, <u>The Development of the Foundations of Mathematical Analysis from Euler to Riemann</u>, M.I.T. Pr., 1970. [1.2.2]

Halmos, Paul R., *The foundations of probability*, <u>Amer. Math. Monthly</u>, 51 (1944) 493-510. [6.1.2]

Hawkins, Thomas, <u>Lebesgue's Theory of Integration: Its Origins and Development</u>, U. of Wisc. Pr., 1970; Chelsea, 1975. [1.2.2]

Kuller, Robert G., *Coin tossing, probability, and the Weierstrass approximation theorem*, <u>Math. Magazine</u>, 37 (1964) 262-265. [6.1.2]

Langer, R.E., *Fourier's series--the genesis and evolution of a theory*, <u>Amer. Math. Monthly</u>, 54 (1947) Suppl. pp. 1-86. [1.2.1]

Myhill, John, *What is a real number?*, <u>Amer. Math. Monthly</u>, 79 (1972) 748-754. [2.3.1]

Steen, Lynn Arthur, *Solving the great bubble mystery*, <u>Science News</u>, 108 (1975) 186-187. [7.1.0]

Stolzenberg, Gabriel *Review of Errett Bishop's Foundations of Constructive Analysis*, <u>Bull. Amer. Math. Soc.</u>, 76 (1970) 301-323. [2.3.1]

Zygmund, Antoni and Fefferman, Charles L., *Fourier analysis*, <u>Encyclopaedia Britannica</u>, 15th ed., 1974, Macropaedia V. 1, pp. 735-757. [4.7.2]

4.4 COMPLEX ANALYSIS

4.4.1 ELEMENTARY

Bers, Lipman, *Complex analysis*, in <u>The Mathematical Sciences--A Collection of Essays for COSRIMS</u>, M.I.T., 1969, pp. 7-20.

> After describing the origin of complex numbers, the author continues in simple language to mention briefly the Riemann hypothesis, the prime number theorem, conformal mapping, the Bieberbach conjecture, the use of conformal mapping in aerodynamics, and the theory of several complex variables. By design, the article whets the appetite rather than satisfies it.

Greene, Francis A., *Complex numbers*, <u>Encyclopedia Americana</u>, 1976, V. 7, pp. 460-461.

Kolata, Gina Bari, *Riemann hypotheses: elusive zeros of the zeta functions*, <u>Science</u>, 185 (1974) 429-431.

> Elementary exposition of various research results motivated by attempts to confirm the Riemann hypothesis. Includes, in particular, a discussion of Levinson's result that one-third of the zeros of the zeta function conform to Riemann's hypothesis and Bombieri's similar confirmation concerning the average distribution of the zeros.

Walsh, J.L., *Complex numbers and complex variables*, <u>McGraw-Hill Encyclopedia of Science and Technology</u>, 1960, V. 3, pp. 336-343.

Walsh, J.L., *Conformal mapping*, <u>McGraw-Hill Encyclopedia of Science and Technology</u>, 1960, V. 3, pp. 395-397.

Wightman, A.S., *Analytic functions and elementary particles*, in <u>The Mathematical Sciences--A Collection of Essays for COSRIMS</u>, M.I.T., 1969, pp. 116-127.

> The author describes briefly how a physicist portrays the results of collisions in elementary particle physics. Then he raises a question concerning the mathematical functions which are used to represent this portrayal, a question which can be paraphrased as follows: granted that only piecewise smooth functions should be used, why do we require the ultimate in smoothness--namely, analyticity? He gives two answers, one in terms of analytic continuation and its physical consequences, the other in terms of the Laplace transform where the finite speed of light guarantees that the transform function is analytic.

4.4.2 ADVANCED

Beckenbach, Edwin F., *Conformal mapping methods*, in E.F. Beckenbach, <u>Modern Mathematics for the Engineer</u>, McGraw-Hill, 1956, pp. 361-388.

    An outline of conformal mapping, from its origins in cartography to the interactions of conformality with complex analysis and potential theory. Includes background knowledge necessary for the construction of conformal maps, namely: existence theory, so as to know what maps are possible; geometric properties, so as to anticipate the maps' behavior; and explicit functions, so as to start with a good approximation.

Connel, E.H., *A classical theorem in complex variables*, <u>Amer. Math. Monthly</u>, 72 (1965) 729-732.

    One of the interesting accomplishments of the field of topological analysis is to bring out the topological heart of the theorem in complex analysis that differentiability in a region implies continuous differentiability (and, of course, more). This article gives a brief and self-contained outline of this result together with the stronger assertion that differentiability implies the existence of the second derivative.

Deavours, C.A., *The quaternion calculus*, <u>Amer. Math. Monthly</u>, 80 (1973) 995-1008.

    A generalization of complex analysis to the quaternionic case, showing how Maxwell's equations can be formulated in this context. The little known subject goes back to work of Fueter in the mid-thirties.

Iwasawa, Kenkichi and Narasimhan, Raghavan, *Complex analysis*, <u>Encyclopaedia Britannica</u>, 15th ed., 1974, Macropaedia V. 1, pp. 719-735.

    A brief outline of classical one-variable theory by the first author is followed by a more extensive discussion of the theory of several complex variables by the second. The latter includes brief mention of a whole series of decidedly graduate-level topics such as Grothendieck's generalization of Hirzebruch's generalization of the Riemann-Roch theorem. The article concludes with a section devoted to potential theory.

Keldyš, M.V., *Functions of a complex variable*, in A.D. Aleksandrov, et al., <u>Mathematics--Its Content, Methods and Meaning</u>, V. 2, M.I.T., 1963, pp. 139-195.

    This survey of functions of a complex variable stresses the applicability of the subject within and without mathematics. Nonetheless, since the article is brief, much of it is devoted to complex function theory proper. The main applications discussed are to hydrodynamics, airfoils, and conformal mapping. The article is a good place for a beginner to get a quick preview or for one who has forgotten his complex function theory to get a quick review.

Lehner, Joseph, *The Picard theorems*, <u>Amer. Math. Monthly</u>, 76 (1969) 1005-1012.

    The "small" Picard theorem states that a function which is analytic in the entire finite complex plane--i.e., an entire function --is either constant or assumes all complex values with at most one exception: it thus substantially strengthens Liouville's theorem that a nonconstant function is unbounded. The "great" Picard theorem makes a similar assertion in the neighborhood of an isolated essential singularity. The author returns to Picard's use of modular functions to prove the theorems, and in the process reviews various aspects of complex analysis.

Nevanlinna, Rolf, *Methods in the theory of integral and meromorphic functions*, <u>J. London Math. Soc.</u>, 41 (1966) 11-28.

    "The development of the theory of entire and meromorphic functions during a century [roughly, 1850-1950] shows a clear tendency towards differential geometrical aspects and methods. However, there remain many interesting and difficult open questions to be investigated by analytical tools..." Nevanlinna gives an historical overview of both of these aspects.

Redheffer, Raymond, *The homotopy theorems of function theory*, <u>Amer. Math. Monthly</u>, 76 (1969) 778-787.

    The author formulates and proves in homotopy form Cauchy's theorem and the monodromy theorem. The paper clarifies some basic aspects of complex analysis.

Whyburn, Gordon T., *Developments in topological analysis*, <u>Fund. Math.</u>, 50 (1962) 305-318.

    Deduction of various classical theorems of complex analysis from topological principles.

Zalcman, Lawrence, *Real proofs of complex theorems (and vice versa)*, <u>Amer. Math. Monthly</u>, 81 (1974) 115-137.

    The author discusses zestfully a series of topics from complex analysis which he would like to see become more "a part of the common culture of professional mathematicians." Various examples of such phenomena as natural boundaries and Jordan curves with positive area suggest a possible article entitled "Examples and counterexamples in complex analysis." The subjects discussed include Morera's theorem and its improvements, the Schwarz reflection principle, extensions of holomorphic functions, "blowing up" the boundary, the Pál-Bohr theorem, Tauberian theorems, and the monodromy theorem. Winner of the Chauvenet Prize and the Lester R. Ford award for expository writing in mathematics.

4.4.3 RESEARCH

Ahlfors, Lars V., *Quasiconformal mappings and their applications*, in T.L. Saaty, Lectures on Modern Mathematics, V. 2, Wiley, 1964, pp. 151-164.

    Quasi-conformal mappings and their relationship to complex function theory (in particular). Interesting side comments and historical remarks make it more readable than most papers dealing with such a technical subject.

Bers, Lipman, *Uniformization moduli and Kleinian groups*, Bull. London Math. Soc., 4 (1972) 257-300.

    Winner of the LeRoy P. Steele prize for distinguished exposition of outstanding research.

Bremermann, H.J., *Several complex variables*, in I.I. Hirschman, Jr., Studies in Real and Complex Analysis, M.A.A., 1965, pp. 3-33.

    A lucid account of the basic concepts and key theorems of several complex variables. (The only hint that the subject is--or, at least, was--of interest in quantum field theory is disposed of with a one-sentence reference to an item in the bibliography.)

Curtiss, J.H., *Faber polynomials and the Faber series*, Amer. Math. Monthly, 78 (1971) 577-596; Correction, 79 (1972) 363.

    For appropriate compact regions of the complex plane there exist sequences of polynomials--the Faber polynomials of the region-- such that every analytic function in the region can be expanded in a convergent infinite series in the Faber polynomials. The paper reviews the history and applications of these Faber series.

Evans, G.C., *Modern methods of analysis in potential theory*, Bull. Amer. Math. Soc., 43 (1937) 481-502.

Walsh, J.L., *History of the Riemann mapping theorem*, Amer. Math. Monthly, 80 (1973) 270-276.

RELATED REFERENCES

Allan G.R., *Some aspects of the theory of commutative Banach algebras and holomorphic functions of several complex variables*, Bull. London Math. Soc., 3 (1971) 1-17. [4.6.3]

Bochner, Salomon, *The rise of functions*, Rice Univ. Studies, 56:2 (1970) 3-21. [1.2.2]

Goldstein, L.J., *A history of the prime number theorem*, <u>Amer. Math. Monthly</u>, 80 (1973) 599-615; Correction, 80 (1973) 1115. [3.3.2]

Hersh, Reuben and Griego, Richard J., *Brownian motion and potential theory*, <u>Scientific American</u>, 220 (March 1969) 66-74, 148. [6.1.1]

Weiss, Guido, *Complex methods in harmonic analysis*, <u>Amer. Math. Monthly</u>, 77 (1970) 465-474. [4.7.3]

4.5 DIFFERENTIAL EQUATIONS

4.5.1 ELEMENTARY

Diaz, J.B., *Differential equation*, McGraw-Hill Encyclopedia of Science and Technology, 1960, V. 4, pp. 125-128.

Griffith, J.S., *Differential equations*, in N.J. Hardiman, Exploring University Mathematics, V. 2, Pergamon, 1968, pp. 99-115.

Kline, Morris, *Differential equations*, Encyclopedia Americana, 1976, V. 9, pp. 109-110.

Lefschetz, Solomon, *Linear and nonlinear oscillations*, in E.F. Beckenbach, Modern Mathematics for the Engineer, McGraw-Hill, 1956, pp. 7-29.

> The author uses a series of elementary examples to illustrate basic notions of oscillatory phenomena, for example the behavior of the solutions of the van der Pol equation (obtained from an electric circuit problem) analyzed in terms of the phase plane. He includes a very brief account of important concepts due to Poincaré such as critical points, limit cycles, and index. The article is supplemented by the statement of some stability results in the companion article by Bellman (ibid., pp. 30-35).

Petrovskiĭ, I.C., *Ordinary differential equations*, in A.D. Aleksandrov, et al., Mathematics--Its Content, Methods and Meaning, V. 1, M.I.T., 1963, pp. 311-356.

4.5.2 ADVANCED

Bernhart, Arthur, *Curves of pursuit*, Scripta Math., 20 (1954) 125-141; 23 (1957) 49-65.

Bernhart, Arthur, *Polygons of pursuit*, Scripta Math., 24 (1959) 23-50.

Bernhart, Arthur, *Curves of general pursuit*, Scripta Math., 24 (1959) 189-206.

Hale, Jack K. and LaSalle, Joseph P., *Differential equations: linearity vs. nonlinearity*, SIAM Review, 5 (1963) 249-272.

> Winner of the Chauvenet Prize for expository writing in mathematics.

[4.5]

Lanczos, Cornelius, *Linear systems in self-adjoint form*, Amer. Math. Monthly, 65 (1958) 665-679.

> Winner of the Chauvenet Prize for expository writing in mathematics.

Lax, Peter D., *Numerical solution of partial differential equations*, Amer. Math. Monthly, 72 (1965) Suppl. pp. 74-84.

> "My aim is to convince [readers]...that the theory of difference equations is...more sophisticated than the corresponding theory of partial differential equations...[because] (1) In order to prove that solutions of a sequence of difference equations converge one needs estimates for difference operators which are analogous to [those in partial differential equations, and] (2) Estimates for difference operators are much harder to derive than the corresponding estimates for differential operators. These contentions will be illustrated on a very simple linear equation..." Winner of the Lester R. Ford award for expository writing in mathematics.

Lax, Peter D., *The formation and decay of shock waves*, Amer. Math. Monthly, 79 (1972) 227-241.

> "Existence and uniqueness of solutions is not the end but merely the beginning of a theory of differential equations. The really interesting questions concern the behavior of solutions." This paper discusses the behavior of solutions of non-linear differential equations representing physical conservation laws, with emphasis as indicated by the title. Some of the material discussed is quite recent; it is made easier by considering relatively simple special cases. Winner of the Lester R. Ford award and the Chauvenet Prize for expository writing in mathematics.

Nirenberg, L., *Partial differential equations with applications in geometry*, in T.L. Saaty, Lectures on Modern Mathematics, V. 2, Wiley, 1964, pp. 1-41.

Reid, W.T., *Anatomy of the ordinary differential equation*, Amer. Math. Monthly, 82 (1975) 971-984.

Schiffer, Menahem M., *Boundary value problems in elliptic partial differential equations*, in E.F. Beckenbach, Modern Mathematics for the Engineer, McGraw-Hill, 1956, pp. 110-144.

> A survey of some aspects of the classical theory of elliptic equations, with emphasis on the role of the Green's function (and to a lesser extent the Neumann function and the kernel function). Heuristic reasoning useful for engineers is invoked where possible. A briefer companion article by R. Courant describes some aspects of hyperbolic equations (ibid., pp. 92-109).

Sneddon, Ian N. and Smale, Stephen, *Differential equations*, Encyclopaedia Britannica, 15th ed., 1974, Macropaedia V. 5, pp. 736-767.

>Many examples of applications to physical problems provide motivation and illustration for this discussion of ordinary and partial differential equations.

Sobolev, S.L. and Ladyzenskaja, O.A., *Partial differential equations*, in A.D. Aleksandrov, et al., Mathematics --Its Content, Methods and Meaning, V. 2, M.I.T., 1963, pp. 3-55.

>A readable standard account of elementary partial differential equations starting with the derivation of the basic types of equations of mathematical physics. Methods of solution include separation of variables, the method of potentials, finite differences, and Galerkin's method. It concludes with a brief discussion of generalized solutions.

Treves, François, *Applications of distributions to PDE theory*, Amer. Math. Monthly, 77 (1970) 241-248.

>The author's object is to convey the impact on partial differential equation theory of Schwartz distributions, together with related "diffusion throughout PDE theory of the methods and results of functional analysis..." His method is to give an important theorem for linear partial differential equations with constant coefficients whose proof illustrates these points.

Weyl, Hermann, *Ramifications, old and new, of the eigenvalue problem*, Bull. Amer. Math. Soc., 56 (1950) 115-139.

4.5.3 RESEARCH

Edmunds, D.E., *Quasilinear second order elliptic and parabolic equations*, Bull. London Math. Soc., 2 (1970) 5-28.

>A survey of recent "definitive" work on the existence of (classical) solutions of boundary and initial value problems for the differential equations of the title, emphasizing the work of Serrin.

Gårding, Lars, *Partial differential equations: problems and uniformization in Cauchy's problem*, in T.L. Saaty, Lectures on Modern Mathematics, V. 2, Wiley, 1964, pp. 129-150.

>Intended for the theoretician, this article gives a survey of developments in the general theory of partial differential operators on differentiable manifolds.

Hermann, Robert, *E. Cartan's geometric theory of partial differential equations*, Advances in Math., 1 (1965) 265-317.

Hermes, Henry, *A survey of recent results in differential equations*, SIAM Review, 15 (1973) 453-468.

Hersh, Reuben, *How to classify differential polynomials*, Amer. Math. Monthly, 80 (1973) 641-654.

> Introduces the basic notions of modern linear differential operators: 'elliptic,' 'hyperbolic,' 'parabolic,' 'hypo-elliptic,' and 'correct,' for equations of arbitrary order.

Papanicolaou, G.C., *Stochastic equations and their applications*, Amer. Math. Monthly, 80 (1973) 526-545.

> An outline of the extensive theory of differential equations whose coefficients and initial data may all be stochastic.

Phillips, Ralph S., *Semigroup methods in the theory of partial differential equations*, in E.F. Beckenbach, Modern Mathematics for the Engineer, 2nd Ser., McGraw-Hill, 1961, pp. 100-132.

> "The semigroup method is in essence an abstract analogue of the Laplace-transform approach to the initial value problem for time-invariant partial differential equations..." The author surveys some of the theory and its applications, with emphasis on his own work on dissipative operators and their applications to hyperbolic systems.

Treves, François, *On local solvability of linear partial differential equations*, Bull. Amer. Math. Soc., 76 (1970) 552-571.

> Winner of the Chauvenet Prize for expository writing in mathematics.

RELATED REFERENCES

Aris, Rutherford, *Chemical kinetics and the ecology of mathematics*, Amer. Scientist, 58 (1970) 419-428. [7.1.1]

Boas, Ralph P., Jr., *Inversion of Fourier and Laplace transforms*, Amer. Math. Monthly, 69 (1962) 955-960. [4.2.2]

Cohen, Hirsh, *Nonlinear diffusion problems*, in A.H. Taub, Studies in Applied Mathematics, M.A.A., 1971, pp. 27-64. [7.1.3]

Hirschman, I.I., Jr. and Widder, D.V., *The Laplace transform, the Stieltjes transform, and their generalizations*, in I.I. Hirschman, Jr., <u>Studies in Real and Complex Analysis</u>, M.A.A., 1965, pp. 67-89. [4.2.2]

Kac, Mark, *Can one hear the shape of a drum?*, <u>Amer. Math. Monthly</u>, 73 (1966) Suppl. pp. 1-23. [7.1.2]

Klamkin, Murray S. and Newman, D.J., *The philosophy and applications of transform theory*, <u>SIAM Review</u>, 3 (1961) 10-36. [7.1.2]

Kolata, Gina Bari, *Cascading bifurcations: the mathematics of chaos*, <u>Science</u>, 189 (1975) 984-985. [7.1.1]

Mackie, A.G., *Some comments on existence and uniqueness theorems in applied mathematics with an application to thin airfoil theory*, <u>SIAM Review</u>, 10 (1968) 196-207. [7.1.3]

May, Robert M., *Biological populations with nonoverlapping generations: stable points, stable cycles, and chaos*, <u>Science</u>, 186 (1974) 645-647. [7.2.2]

May, Robert M., *On relationships among various types of population models*, <u>Amer. Naturalist</u>, 107 (1973) 46-57. [7.2.2]

McHugh, James A.M., *An historical survey of ordinary linear differential equations with a large parameter and turning points*, <u>Arch. Hist. Exact Sci.</u>, 7 (1971) 277-324. [1.2.2]

Morse, Marston, *What is analysis in the large?*, <u>Amer. Math. Monthly</u>, 49 (1942) 358-364; also in S.S. Chern, <u>Studies in Global Geometry and Analysis</u>, M.A.A., 1967, pp. 5-15. [5.6.2]

Nitsche, Johannes C.C., *Plateau's problems and their modern ramifications*, <u>Amer. Math. Monthly</u>, 81 (1974) 945-968. [5.3.2]

Schwartz, Laurent, *Some applications of the theory of distributions*, in T.L. Saaty, <u>Lectures on Modern Mathematics</u>, V. 1, Wiley, 1963, pp. 23-58. [4.6.3]

Slater, J.C., *Physics and the wave equation*, Bull. Amer. Math. Soc., 52 (1946) 392-400. [7.1.2]

Smale, Stephen, *Differentiable dynamical systems*, Bull. Amer. Math. Soc., 73 (1967) 747-817. [5.6.3]

Smale, Stephen, *What is global analysis?*, Amer. Math. Monthly, 76 (1969) 4-9. [5.6.2]

Varga, Richard S., *Iterative methods for solving matrix equations*, Amer. Math. Monthly, 72 (1965) Suppl. pp. 67-74. [6.4.2]

4.6 FUNCTIONAL ANALYSIS

4.6.1 ELEMENTARY

Gel'fand, I.M., *Functional analysis*, in A.D. Aleksandrov, et al., Mathematics--Its Content, Methods and Meaning, V. 3, M.I.T., 1963, pp. 227-261.

> An excellent development of the elementary notions and definitions of Hilbert space and operator theory with constant attention to the source of these notions in classical analysis and applied mathematics.

Schwartz, Jacob T., *Functional analysis*, in The Mathematical Sciences--A Collection of Essays for COSRIMS, M.I.T., 1969, pp. 72-83.

> "By generalizing an idea, one can carry over the insight it contains from its area of origin to other areas, perhaps even to [those which] appear unrelated... Thus, what started as a limited idea can grow and spread and unify important concepts in many diverse fields." The author uses this point of view to discuss the roles of the Pythagorean theorem and fixed point theorems in functional analysis. He also enunciates what he calls a "force of context" principle, illustrated, e.g., by the role of the path of a light ray in an optical medium in minimizing time among a family of paths.

4.6.2 ADVANCED

Ficken, F.A., *Some uses of linear spaces in analysis*, Amer. Math. Monthly, 66 (1959) 259-275.

> Expository account beginning with Picard's theorem for $y' = f(x,y)$, then Banach spaces, then Hilbert space. Good bibliography.

Gelbaum, B.R., *Banach algebras and their applications*, Amer. Math. Monthly, 71 (1964) 248-256.

> A quick review of Banach algebras, with some emphasis on topics in classical analysis which motivate interest in these algebras. Six or seven basic theorems are stated, mostly without proof, but there is an outline of the author's elegant proof of a theorem of von Neumann concerning normal operators on a Hilbert space.

Goffman, Casper, *Preliminaries to functional analysis*, in R.C. Buck, Studies in Modern Analysis, M.A.A., 1962, pp. 138-180.

> A survey of such topics as fixed point theorems, compactness arguments, Hilbert spaces, Banach spaces, and Banach algebras with

emphasis on special cases and relations to problems of classical analysis.

Graves, Lawrence M., *What is a functional?*, Amer. Math. Monthly, 55 (1948) 467-472.

Halmos, Paul R., *A glimpse into Hilbert space*, in T.L. Saaty, Lectures on Modern Mathematics, V. 1, Wiley, 1963, pp. 1-22.

Horváth, John, *An introduction to distributions*, Amer. Math. Monthly, 77 (1970) 227-240.

> Distributions are often described as the mathematics which makes the Dirac delta function rigorous. They are vastly more than that: witness, for example, their profound impact on the theory of partial differential equations in the last twenty years. This brief survey covers various aspects of distributions and includes interesting historical remarks. The author's 1966 book on the subject would be an excellent source for further reading.

Hyers, D.H., *Linear topological spaces*, Bull. Amer. Math. Soc., 51 (1945) 1-21.

> Historical survey with many references.

Lorch, Edgar R., *The spectral theorem*, in R.C. Buck, Studies in Modern Analysis, M.A.A., 1962, pp. 88-137.

> Starting with the definition of a linear transformation on three-dimensional Euclidean space, the author proceeds (by what he terms a "gentle" approach) to various cases of the spectral theorem for infinite dimensional Hilbert space. While there is brief mention of the historical role of integral equations, the emphasis is on the statement and motivation of the spectral theorem itself rather than on its applications.

Mackey, George W., *Operator theory*, McGraw-Hill Encyclopedia of Science and Technology, 1960, V. 9, pp. 338-341.

McShane, E.J., *Trends in analysis*, Amer. Math. Monthly, 74 (1967) Suppl. pp. 65-79.

> The author sets out to explain the motivation, the advantages, and the drawbacks of the introduction of abstract methods into analysis, illustrating his points with a series of historically structured examples. Example: von Neumann's introduction of abstract Hilbert space gave a common setting for the quantum theories of Schrödinger and Heisenberg. Example: Mikusinski's operational calculus has advantages over the Laplace transform in explicating the Heaviside calculus. Other examples are drawn from distribution theory and probability theory.

Packel, Edward W., *Hilbert space operators and quantum mechanics*, Amer. Math. Monthly, 81 (1974) 863-873.

   An article well suited to the reader who knows the rudiments of Hilbert space theory and would like to get a glimpse of how this theory can be used to formulate some of the concepts and results of quantum mechanics--for example the Heisenberg uncertainty principle. The quantum mechanics is sketchy (Schrödinger's equation is not mentioned), but this is consistent with the author's goal "to elucidate in an elementary and necessarily simplified fashion this important physical role of Hilbert space."

Schwartz, Jacob T., Morrey, Charles B., Jr. and Treves, François, *Functional analysis*, Encyclopaedia Britannica, 15th ed., 1974, Macropaedia V. 1, pp. 757-772.

   This article is divided into three parts: a broad view of functional analysis by Schwartz; calculus of variations by Morrey; and generalized functions, or theory of distributions, by Treves.

Steen, Lynn Arthur, *Highlights in the history of spectral theory*, Amer. Math. Monthly, 80 (1973) 359-381.

   The author traces the evolution of spectral theory, from early beginnings to the Gelfand-Naimark theorem. There is natural emphasis on the work of Hilbert and his followers, the interaction with quantum mechanics, and the co-extensive work of von Neumann and Stone. In the process he shows not only the interrelationships of various contributions within spectral theory, but also how the emerging spectral theory interacted with the development of the Lebesgue integral on the one hand, and that of quantum mechanics on the other. Winner of the Lester R. Ford award for expository writing.

Temple, G., *Theories and applications of generalized functions*, J. London Math. Soc., 28 (1953) 134-148.

4.6.3 RESEARCH

Allan, G.R., *Some aspects of the theory of commutative Banach algebras and holomorphic functions of several complex variables*, Bull. London Math. Soc., 3 (1971) 1-17.

Bonsall, F.F., *A survey of Banach algebra theory*, Bull. London Math. Soc., 2 (1970) 257-274.

   "The theory of Banach algebras began in 1939 with the appearance of the first of Gelfand's brilliant series of papers on 'normed rings.' ...In the middle fifties the subject seemed to lose its momentum and was deserted by some of its leading exponents. I shall try to show that the subject has been restored to full vigor in recent years... I shall sketch the classical theory, give a somewhat fuller account of some recent work, and end with some guesses about directions in which progress looks possible."

De Jager, E.M., *Theory of distributions*, in E. Roubine, Mathematics Applied to Physics, Springer-Verlag, 1970, pp. 52-110.

> A broad, self-contained introduction written for a UNESCO conference aimed at improving the teaching of physics.

Fillmore, P.A., *The shift operator*, Amer. Math. Monthly, 81 (1974) 717-723.

> "The shift operator has been known for many years, at first as an interesting example, but more recently as a fundamental building block in the structure theory of operators on Hilbert space."

Graves, Lawrence M., *Nonlinear mappings between Banach spaces*, in I.I. Hirschman, Jr., Studies in Real and Complex Analysis, M.A.A., 1965, pp. 34-54.

> Definitions, examples and theory of linear and nonlinear mappings between Banach spaces with emphasis on implicit function theorems. Closes with some applications to the theory of differential equations.

Halmos, Paul R., *What does the spectral theorem say?*, Amer. Math. Monthly, 70 (1963) 241-247.

> The author's aim is not to give a proof of the spectral theorem, although he does outline one. Rather, he is interested in presenting a formulation of the spectral theorem (for a Hermitian operator on a complex Hilbert space) that will be easy to grasp: every Hermitian operator is unitarily equivalent to a multiplication (by a bounded, real-valued function on a suitable measure space).

Hille, Einar, *Topics in classical analysis*, in T.L. Saaty, Lectures on Modern Mathematics, V. 3, Wiley, 1965, pp. 1-57.

> The "topics" discussed are functional inequalities, functional equations, mean values, transfinite diameters, and potential theory.

Hille, Einar, *What is a semi-group?*, in I.I. Hirschman, Jr., Studies in Real and Complex Analysis, M.A.A., 1965, pp. 55-66.

> In this brief account of semi-groups of operators, the author first illustrates some of the principles of the subject by restricting himself to the case of two-by-two matrices. After mentioning six mathematical uses of one parameter semi-groups, he closes with the statement of the Hille-Yosida theorem and mention of its relationship to the abstract Cauchy problem.

Lorch, Edgar R., *The structure of normed Abelian rings*, Bull. Amer. Math. Soc., 50 (1944) 447-463.

> Extensive survey with lots of examples.

McKelvey, Robert, *Symmetric differential operators*, <u>Amer. Math. Monthly</u>, 71 (1964) 119-129.

> An account of the theory of symmetric differential operators leading to a description of Naimark's theory of extension into transcendent Hilbert spaces. Questions of extension are cast in a "particularly simple form" in terms of boundary conditions, spectral matrices, and eigen-function expansions.

Royden, H.L., *Function algebras*, <u>Bull. Amer. Math. Soc.</u>, 69 (1963) 281-298.

> Lots of examples, good bibliography.

Schwartz, Laurent, *Some applications of the theory of distributions*, in T.L. Saaty, <u>Lectures on Modern Mathematics</u>, V. 1, Wiley, 1963, pp. 23-58.

> "In this paper we give a summary of the main results of the theory of distributions, and some selected applications." The applications referred to are within mathematics, although since they pertain to partial differential equations, they hint at possible outside applications. However, this exposition is aimed exclusively at mathematicians.

Taylor, Angus E., *Notes on the history of the uses of analyticity in operator theory*, <u>Amer. Math. Monthly</u>, 78 (1971) 331-342.

> After recalling that Fredholm's equation of the second kind has a solution which involves analytic functions of the parameter, the author traces the generalization of the notion of analyticity to operator-valued functions as it appears, for example, in the work of Riesz, Dunford's theory of spectral operators, Taylor's own work, and elsewhere.

Wermer, John, *Banach algebras and analytic functions*, <u>Advances in Math.</u>, 1 (1965) 51-102.

Wermer, John, *Uniform approximation and maximal ideal spaces*, <u>Bull. Amer. Math. Soc.</u>, 68 (1962) 298-305.

RELATED REFERENCES

Bernkopf, Michael, *A history of infinite matrices*, <u>Arch. Hist. Exact Sci.</u>, 4 (1968) 308-358.   [1.2.3]

Bernkopf, Michael, *The development of function spaces with particular reference to their origins in integral equation theory*, <u>Arch. Hist. Exact Sci.</u>, 3 (1966) 1-96.   [1.2.3]

Bernstein, Allan R., *Non-standard analysis*, [2.4.3]
in M.D. Morley, Studies in Model Theory, M.A.A.,
1973, pp. 35-58.

Halmos, Paul R., *Finite-dimensional Hilbert* [3.4.2]
spaces, Amer. Math. Monthly, 77 (1970) 457-
464.

Mackey, George W., *Quantum mechanics and* [7.1.2]
Hilbert space, Amer. Math. Monthly, 64 (1957)
Suppl. pp. 45-57.

Monna, A.F., Functional Analysis in Historical [1.2.3]
Perspective, Halsted Pr., 1973.

Treves, François, *Applications of distribu-* [4.5.2]
tions to PDE theory, Amer. Math. Monthly, 77
(1970) 241-248.

4.7 Special Topics

4.7.1 ELEMENTARY

Hestenes, Magnus R., *An elementary introduction to the calculus of variations*, Math. Magazine, 23 (1950) 249-267.

Krylov, V.I., *The calculus of variations*, in A.D. Aleksandrov, et al., Mathematics--Its Content, Methods and Meaning, V. 2, M.I.T., 1963, pp. 119-138.

> A brief, self-contained account of the classical calculus of variations, with some mention of its applicability in mechanics.

4.7.2 ADVANCED

Birkhoff, George D., *What is the ergodic theorem?*, Amer. Math. Monthly, 49 (1942) 222-226.

Hestenes, Magnus R., *Elements of the calculus of variations*, in E.F. Beckenbach, Modern Mathematics for the Engineer, McGraw-Hill, 1956, pp. 59-91.

> The author summarizes the basic notions and some of the highpoints of the classical calculus of variations, including the brachistochrone problem, a minimum area problem, geodesics, Hamilton's principle and the Hamiltonian, and isoperimetric problems.

Jones, D.S., *Asymptotic behavior of integrals*, SIAM Review, 14 (1972) 286-317.

Miles, John W., *Integral transforms*, in E.F. Beckenbach, Modern Mathematics for the Engineer, 2nd Ser., McGraw-Hill, 1961, pp. 68-99.

> A brief survey of standard integral transforms (Fourier, Laplace, Hankel, Mellin, etc.) and their engineering applications.

Pólya, George, *Circle, sphere, symmetrization and some classical physical problems*, in E.F. Beckenbach, Modern Mathematics for the Engineer, 2nd Ser., McGraw-Hill, 1961, pp. 420-441.

> The author describes the treatment of various isoperimetric type problems by Steiner symmetrization. He then considers properties of the lowest eigenvalue of a vibrating membrane, using the same techniques (and with the same results: the figure which minimizes the perimeter also minimizes the eigenvalue). A high point is the way the area formula for a surface is combined with symmetrization to estimate Dirichlet integrals. In the final sections uniqueness problems are treated by "alternative symmetrization."

Weiss, Guido, *Harmonic analysis*, in I.I. Hirschman, Jr., Studies in Real and Complex Analysis, M.A.A., 1965, pp. 124-178.

This article serves as a "halfway house" between what is taught in advanced calculus concerning Fourier series (and perhaps Fourier integrals), and what a textbook on Fourier series and integrals would present. Three pages near the end contain a concise description of what harmonic analysis on locally compact abelian groups is about and its relationship to traditional Fourier series. Winner of the Chauvenet Prize for expository writing in mathematics.

Zygmund, Antoni and Fefferman, Charles L., *Fourier analysis*, Encyclopaedia Britannica, 15th ed., 1974, Macropaedia V. 1, pp. 735-757.

The first section gives background material on basic notions of infinite series. This is followed by sections entitled Fourier series, harmonic analysis and integral transforms, and representations of groups and algebras.

4.7.3 RESEARCH

Cameron, R.H., *Some introductory exercises in the manipulation of Fourier transforms*, National Math. Magazine, 15 (1941) 331-356.

Winner of the Chauvenet Prize for expository writing in mathematics.

Coifman, R.R. and Weiss, Guido, *Representations of compact groups and spherical harmonics*, L'Enseignement Math., 14 (1968) 121-173.

Fefferman, Charles L., *Recent progress in classical Fourier analysis*, in Proc. Inter. Cong. Math. (1974), V. 1, Canad. Cong. Math., 1975, pp. 95-118.

This survey is rather technical, but it does an excellent job of tying together a number of important topics. After brief comments on the one variable case as it existed about 1950 (with emphasis on the Littlewood-Paley theorem), the author moves on to several variables with natural emphasis on singular integrals, pseudodifferential operators, etc., as basic tools. Wherever possible the intuitive ideas are emphasized, e.g., in the use of various gambling situations from probability theory which are relevant to the development.

Fink, A.M., *Almost periodic functions invented for specific purposes*, SIAM Review, 14 (1972) 572-581.

A concise summary of various results in the theory of almost periodic functions (due to Bohr, Bochner, von Neumann, Markoff, Favard, Amerio and Seifert) together with sketched proofs of the equivalence of different definitions of almost periodicity. The theme of the article is the interaction of this material with the study of almost periodic solutions of differential equations.

Halmos, Paul R., *Recent progress in ergodic theory*, Bull. Amer. Math. Soc., 67 (1961) 70-80.

An unusually intuitive treatment with few formulas.

Keller, Joseph B. and McLaughlin, D.W., *The Feynman integral*, Amer. Math. Monthly, 82 (1975) 451-465.

Mackey, George W., *Ergodic theory and its significance for statistical mechanics and probability theory*, Advances in Math., 12 (1974) 178-286.

Winner of the LeRoy P. Steele prize for distinguished exposition of outstanding research.

Mackey, George W., *Functions on locally compact groups*, Bull. Amer. Math. Soc., 56 (1950) 385-412.

Morse, Marston, *Trends in analysis*, J. Franklin Inst., 251 (1951) 33-43.

Ornstein, D.S., *Measure-preserving transformations and random processes*, Amer. Math. Monthly, 78 (1971) 833-840.

Ergodic theory was given new impetus by the introduction of the invariant "entropy" by Kolmogoroff in 1958 (motivated by Shannon's work on information theory). An immediate application was to show that not all Bernoulli shifts are isomorphic--those with different entropies cannot be. Ornstein has now proved the converse: Bernoulli shifts with the same entropy are isomorphic. This expository paper gives three different characterizations of the notion of Bernoulli shift, describes why Bernoulli shifts are important, and discusses their relationship to ergodic theory and probability theory, including various results due to the author.

Rosser, J. Barkley, *Asymptotic formulas and series*, in E.F. Beckenbach, Modern Mathematics for the Engineer, 2nd Ser., McGraw-Hill, 1961, pp. 133-163.

A discussion of various aspects of asymptotic methods that are dealt with more deeply in standard advanced treatises. The idea is to give the reader a feel for what the subject is all about.

Seeley, Robert T., *Spherical harmonics*, <u>Amer. Math. Monthly</u>, 73 (1966) Suppl. pp. 115-121.

"The object of the present article is to give a concise and elementary exposition of spherical harmonics. ...By elementary, we mean independent of any knowledge of special functions."

Segel, Lee A., *The importance of asymptotic analysis in applied mathematics*, <u>Amer. Math. Monthly</u>, 73 (1966) 7-14.

Taylor, Joseph L., *Measure algebras*, <u>CBMS Reg. Conf. Ser. in Math.</u>, No. 16, Amer. Math. Soc., 1972.

Winner of the LeRoy P. Steele prize for distinguished exposition of outstanding research.

Weiss, Guido, *Complex methods in harmonic analysis*, <u>Amer. Math. Monthly</u>, 77 (1970) 465-474.

The author quotes Zygmund's description of the classical theory of Fourier series "as the meeting ground of real and complex variables," and illustrates it with four applications of complex variables to harmonic analysis. A key tool is Fatou's theorem that a bounded analytic function on the open unit disc has radial limits almost everywhere. Weiss then describes how complex methods can also be applied to harmonic analysis on higher dimensional Euclidean spaces.

Wiener, Norbert, *The theory of prediction*, in E.F. Beckenbach, <u>Modern Mathematics for the Engineer</u>, McGraw-Hill, 1956, pp. 165-190.

The author gives a no-nonsense survey of his theory of prediction. The level of presentation is set right at the start with his sketch of proofs of the ergodic theorem of Koopman and von Neumann (author's terminology) and the ergodic theorem of Birkhoff.

Williamson, J.H., *Harmonic analysis on semigroups*, <u>J. London Math. Soc.</u>, 42 (1967) 1-41.

A survey of methods in semigroups that are inspired by harmonic analysis, a field begun by Hewitt in 1955.

RELATED REFERENCES

Payne, Lawrence F., *Isoperimetric inequalities and their applications*, <u>SIAM Review</u>, 9 (1967) 453-488.   [6.3.3]

Steen, Lynn Arthur, *Order from chaos*, <u>Science News</u>, 107 (1975) 292-293.   [6.1.0]

5.1 Classical Geometry

5.1.0 GENERAL

Archibald, R.C., *The first translation of Euclid's Elements into English and its source*, Amer. Math. Monthly, 57 (1950) 443-452.

Barnard, Raymond W., et al., *Analytic and trigonometric geometry*, Encyclopaedia Britannica, 15th ed., 1974, Macropaedia V. 7, pp. 1076-1093.

> A review of advanced high school mathematics. Some readers will be interested in the collection of special curves (illustrated); others may be interested in the one page summary of spherical trigonometry.

Blumenthal, Leonard M., *Analytic geometry*, McGraw-Hill Encyclopedia of Science and Technology, 1960, V. 1, pp. 387-392.

Coxeter, H.S. MacDonald, *Non-Euclidean geometry*, in The Mathematical Sciences--A Collection of Essays for COSRIMS, M.I.T., 1969, pp. 52-59.

> Without going into detail, Coxeter highlights some key ideas in the development of non-Euclidean geometries, thus keeping the account accessible to a large audience.

Dudley, Underwood, *Who was the first non-Euclidean?*, Math. Spectrum, 6 (1973-74) 41-46.

> Argues that Lobachevsky and Bolyai got their ideas indirectly from Gauss.

Hahn, Hans, *Geometry and intuition*, Scientific American, 190 (April 1954) 84-91, 108; also in M. Kline, Mathematics in the Modern World, Freeman, 1968, pp. 184-188, 399.

Holden, Alan, Shapes, Space and Symmetry, Columbia U. Pr., 1971.

Knebelman, M.S., *Graphical coordinate systems*, McGraw-Hill Encyclopedia of Science and Technology, 1960, V. 3, pp. 454-456.

Lieber, Lillian R., Non-Euclidean Geometry or Three Moons in Mathesis, Science Pr., 1940.

> The writings of Lillian Lieber are the nursery rhymes of the literature of mathematics. They have a unique charm which can only be conveyed by an actual reading.

Rogers, Pat, *The parallel axiom*, Math. Spectrum, 5 (1972-73) 58-66.

Examines criteria for an axiom system, Euclid's deficiencies, and Hilbert's axioms; and gives Poincaré's model for Lobachevskian geometry.

Stevens, Peter S., Patterns in Nature, Atlantic-Little, Brown, 1974.

A beautiful, essentially nonmathematical approach to geometric patterns: flow, branching, spirals, packing, soap films.

Tóth, Imre, *Non-Euclidean geometry before Euclid*, Scientific American, 221 (November 1969) 87-98, 166.

Zassenhaus, Hans J., *What is an angle?*, Amer. Math. Monthly, 61 (1954) 369-378.

Axiomatic, elementary; no bibliography.

5.1.1 ELEMENTARY

Aleksandrov, A.D., *Non-euclidean geometry*, in A.D. Aleksandrov, et al., Mathematics--Its Content, Methods and Meaning, V. 3, M.I.T., 1963, pp. 97-189.

Allendoerfer, Carl B., *Generalizations of theorems about triangles*, Math. Magazine, 38 (1965) 253-259.

Winner of the Lester R. Ford award for expository writing.

Busemann, Herbert, *Non-Euclidean geometry*, Math. Magazine, 24 (1950) 19-34.

Coxeter, H.S. MacDonald, *A geometrical background for de Sitter's world*, Amer. Math. Monthly, 50 (1943) 217-228.

Coxeter, H.S. MacDonald, *Non-Euclidean geometry*, Encyclopaedia Britannica, 15th ed., 1974, Macropaedia V. 7, pp. 1112-1120.

This is a particularly readable exposition of various models of the hyperbolic plane (Beltrami, Klein, Poincaré, Liebmann). Coordinates and transformations are also discussed.

Coxeter, H.S. MacDonald, *The problem of Apollonius*, Amer. Math. Monthly, 75 (1968) 5-15.

Coxeter, H.S. MacDonald and Greitzer, S.L., Geometry Revisited, New Math. Libr., No. 19, Random House, 1967; M.A.A., 1975.

Classical Geometry [5.1]

Du Val, Patrick, *Projective geometry*, Encyclopaedia Britannica, 15th ed., 1974, Macropaedia, V. 7, pp. 1120-1125.

Covers such topics as projection, homogeneous coordinates, and complex geometry. The final section discusses the abstract geometries obtained by choosing coordinates from various algebraic systems, and the problem of their classification.

Fejes Toth, L., *What the bees know and what they do not know*, Bull. Amer. Math. Soc., 70 (1964) 468-481.

Discussion of geometry of honeycombs and various generalizations.

Forder, Henry G. and Valentine, Frederick A., *Euclidean geometry*, Encyclopaedia Britannica, 15th ed., 1974, Macropaedia V. 7, pp. 1099-1112.

A review of Euclid is followed by a more abstract discussion based on Hilbert's version of the axioms. Other topics included are transformation geometry, constructions, and convexity.

Gans, David, *An introduction to elliptic geometry*, Amer. Math. Monthly, 62 (1955) Suppl. pp. 66-75.

Golos, Ellery B., *Projective geometry*, Encyclopedia Americana, 1976, V. 12, pp. 487-492.

Kay, David C., *Non-Euclidean geometry*, Encyclopedia Americana, 1976, V. 12, pp. 494-499.

Knebelman, M.S., et al., *Geometry*, McGraw-Hill Encyclopedia of Science and Technology, 1960, V. 6, pp. 150-164.

Lanczos, Cornelius, Space Through the Ages, Academic Pr., 1970.

Lyon, Thoburn C., *Projection*, Encyclopedia Americana, 1976, V. 22, pp. 644-650b.

Niven, Ivan M. and Zuckerman, H.S., *Lattice points and polygonal area*, Amer. Math. Monthly, 74 (1967) 1195-1200.

A proof of a specialized but interesting result often called Pick's theorem: the area of a simple polygon all of whose vertices are lattice points is equal to the number of interior lattice points, plus half the lattice points on the boundary, minus one.

Prenowitz, Walter, *A contemporary approach to classical geometry*, Amer. Math. Monthly, 68 (1961) Suppl. pp. 1-67.

Struik, Dirk J., *Descriptive geometry*, Encyclopedia Americana, 1976, V. 12, pp. 492-494.

Thurston, Hugh, *Tangents: an elementary survey*, <u>Math. Magazine</u>, 42 (1969) 1-11.

Willoughby, Stephen S., *Analytic geometry*, <u>Encyclopedia Americana</u>, 1976, V. 12, pp. 483-487.

Willoughby, Stephen S., *Euclidean geometry*, <u>Encyclopedia Americana</u>, 1976, V. 12, pp. 479-483.

Yaglom, I.M., <u>Geometric Transformations</u>, New Math. Libr., No. 8, Random House, 1962; M.A.A., 1975.

Yaglom, I.M., <u>Geometric Transformations II</u>, New Math. Libr., No. 21, Random House, 1968; M.A.A., 1975.

Yaglom, I.M., <u>Geometric Transformations III</u>, New Math. Libr., No. 24, Random House, 1973; M.A.A., 1975.

5.1.2 ADVANCED

Bruck, R.H., *Recent advances in the foundations of Euclidean plane geometry*, <u>Amer. Math. Monthly</u>, 62 (1955) Suppl. pp. 2-17.
  Winner of the Chauvenet Prize for expository writing in mathematics.

Coxeter, H.S. MacDonald, *Geometry*, in T.L. Saaty, <u>Lectures on Modern Mathematics</u>, V. 3, Wiley, 1965, pp. 58-94.
  "...Seven silver dollars can be placed on a table so that one is surrounded by a ring of six." The analogous problem in 3-space is to find the maximum number M_3 of, say, basketballs which will touch a given basketball. "In 1694 Isaac Newton told David Gregory that $M_3 = 12$, but Gregory asserted that $M_3 = 13$. R. Hoppe, 180 years later, proved that Newton's value is correct." Coxeter relates this problem to the Gaussian quadratic forms of a lattice and to certain Lie algebras. The paper is divided into a series of such two or three page vignettes on geometric subjects: Kakeya problems in 2-space and 3-space, ordered geometry, integral quaternions and integral octaves, projective geometry, hyperbolic geometry, etc.

Pedoe, Daniel, *On a theorem in geometry*, <u>Amer. Math. Monthly</u>, 74 (1967) 627-640.
  Winner of the Lester R. Ford award for expository writing.

Scherk, Peter, *Some concepts of conformal geometry*, <u>Amer. Math. Monthly</u>, 67 (1960) 1-30.
  Linear algebra and non-Euclidean geometry.

Classical Geometry   *129*   [5.1]

RELATED REFERENCES

Aaboe, Asger, Episodes From the Early History of Mathematics, New Math. Libr., No. 13, Random House, 1964; M.A.A., 1975.   [1.2.1]

Carter, F.L., *Perspective drawing by numbers*, Math. Gazette, 53 (1969) 133-139.   [7.6.1]

Chakerian, G.D. and Lange, L.H., *Geometric extremum problem*, Math. Magazine, 44 (1971) 57-69.   [6.3.1]

Chittenden, J. Brace, *Quadrature of the circle*, Encyclopedia Americana, 1976, V. 23, pp. 52-53.   [1.2.1]

Coxeter, H.S. MacDonald, *The space-time continuum*, Historia Math., 2 (1975) 289-298.   [1.2.1]

Crowe, Donald W., *The geometry of African art I. Bakuba art*, J. of Geometry, 1 (1971) 169-182; *II. A catalog of Benin patterns*, Historia Math., 2 (1975) 253-271.   [7.6.1]

Cunningham, F., *The Kakeya problem for simply connected and for star-shaped sets*, Amer. Math. Monthly, 78 (1971) 114-129.   [6.3.2]

Flegg, Graham, From Geometry to Topology, Crane Russak, 1974.   [5.5.1]

Friedrichs, K.O., From Pythagoras to Einstein, New Math. Libr., No. 16, Random House, 1965; M.A.A., 1975.   [7.1.1]

Goldstein, Marie, *The historical development of group theoretical ideas in connection with Euclid's axiom of congruence*, Notre Dame J. of Formal Logic, 13 (1972) 331-349.   [1.2.2]

Grünbaum, Adolf, *Geometry, chronometry and empiricism*, in H. Feigel and G. Maxwell, Minnesota Studies in the Philosophy of Science, V. 3, U. Minn. Pr., 1962, pp. 405-526.   [2.2.1]

Grünbaum, Adolf, *Reply to Hilary Putnam's 'An examination of Grünbaum's philosophy of geometry'*, in R.S. Cohen and M.W. Wartofsky, Boston Studies in the Philosophy of Science, V. 5, D. Reidel, 1969, pp. 1-150.   [2.2.1]

Hilbert, David and Cohn-Vossen, Stephan, [5.5.1]
Geometry and the Imagination, Chelsea, 1952.

Honsberger, Ross, Ingenuity in Mathematics, [3.1.1]
New Math. Libr., No. 23, Random House, 1970;
M.A.A., 1975.

Kazarinoff, Nicholas D., Geometric Inequali- [6.3.0]
ties, New Math. Libr., No. 4, Random House,
1961; M.A.A., 1975.

Kline, Morris, *Geometry: history and develop-* [1.2.1]
ment, Encyclopedia Americana, 1976, V. 12, pp.
471-478.

Kuhn, Harold W., *"Steiner's" problem revisited*, [6.3.2]
in G.B. Dantzig and B.C. Eaves, Studies in
Optimization, M.A.A., 1974, pp. 52-70.

Mahoney, Michael S., *Another look at Greek* [1.2.0]
geometrical analysis, Arch. Hist. Exact Sci.,
5 (1968) 318-348.

Nevanlinna, Rolf, *Reform in teaching mathema-* [1.6.0]
tics, Amer. Math. Monthly, 73 (1966) 451-464.

Polachek, Harry, *The structure of the honey-* [7.2.1]
comb, Scripta Math., 7 (1940) 87-98.

Pólya, George, *Circle, sphere, symmetrization* [4.7.2]
and some classical physical problems, in E.F.
Beckenbach, Modern Mathematics for the Engineer, 2nd Ser., McGraw-Hill, 1961, pp. 420-441.

Seidenberg, A., *On the area of a semi-circle*, [1.2.0]
Arch. Hist. Exact Sci., 9 (1972) 171-211.

Stein, Sherman K., *Algebraic tiling*, Amer. [3.5.2]
Math. Monthly, 81 (1974) 445-462.

Zippin, Leo, Uses of Infinity, New Math. [2.1.0]
Libr., No. 7, Random House, 1962; M.A.A., 1975.

5.2 COMBINATORIAL GEOMETRY

5.2.0 GENERAL

Kelly, Paul J., *Plane convex figures*, <u>NCTM Twenty-Eighth Yearbook</u>, 1963, 251-264.

   Discussion of curves of constant width.

5.2.1 ELEMENTARY

Klee, Victor, *What is a convex set?*, <u>Amer. Math. Monthly</u>, 78 (1971) 616-631.

   Based on an encyclopedia article by the author, this paper highlights various aspects of convexity such as packing problems, geometric probability, intersection properties, the combinatorial study of polyhedra, and the role of convexity in topological vector spaces. Large bibliography. Winner of the Lester R. Ford award for expository writing.

5.2.2 ADVANCED

Bose, Raj C. and Grünbaum, Branko, *Combinatorics and combinatorial geometry*, <u>Encyclopaedia Britannica</u>, 15th ed., 1974, Macropaedia V. 4, pp. 942-954.

   The article includes a discussion of some historically important problems in combinatorial geometry and examples of frequently used methods.

Chakerian, G.D., *Intersection and covering properties of convex sets*, <u>Amer. Math. Monthly</u>, 76 (1969) 753-766.

   A survey of certain parts of convexity theory, with emphasis on Helly-type theorems. Helly's theorem, stated for ordinary 3-space, asserts that if a finite collection of compact, convex subsets has the property that every four (or fewer) members have a point in common, then there must be a point which is common to all members of the collection. This theorem and its many variants play an important role in convexity theory.

Valentine, Frederick A., *Visible shorelines*, <u>Amer. Math. Monthly</u>, 77 (1970) 146-152.

   Description of several Helly-type theorems.

5.2.3 RESEARCH

Grünbaum, Branko and Shephard, G.C., *Convex polytopes*, Bull. London Math. Soc., 1 (1969) 257-300.

A survey of convex polytopes of arbitrary dimension, listing main results and unsolved problems.

Klee, Victor, *The Euler characteristic in combinatorial geometry*, Amer. Math. Monthly, 70 (1963) 119-127.

Zassenhaus, Hans J., *Modern developments in the geometry of numbers*, Bull. Amer. Math. Soc., 67 (1961) 427-439.

Packing problems and covering problems. A good general introduction; assumes no prior knowledge.

RELATED REFERENCES

Birkhoff, George D., Aesthetic Measure, Harvard U. Pr., 1933. [1.7.0]

Grünbaum, Branko, *Polygons in arrangements generated by n points*, Math. Magazine, 46 (1973) 113-119. [3.1.1]

Honsberger, Ross, Mathematical Gems, M.A.A., 1973. [3.1.1]

5.3 Differential Geometry

5.3.1 ELEMENTARY

Aleksandrov, A.D., *Curves and surfaces*, in A.D. Aleksandrov, et al., <u>Mathematics--Its Content, Methods and Meaning</u>, V. 2, M.I.T., 1963, pp. 57-117.

> A brief, self-contained description of the differential geometry of curves and surfaces in three-space, with emphasis on the case of surfaces (as evidenced by the omission of the Frenet formulas from the discussion of curves). It therefore has a lot of overlap with many standard textbooks, but it has the advantage that the author devotes considerable effort to getting ideas across with the aid of diagrams, analogies, and discussion.

Courant, Richard, *Soap film experiments with minimal surfaces*, <u>Amer. Math. Monthly</u>, 47 (1940) 167-174.

Douglas, Jesse, *The problem of Plateau*, <u>Scripta Math.</u>, 5 (1938) 159-164.

Eggleston, H.G., *The isoperimetric problem*, in N.J. Hardiman, <u>Exploring University Mathematics</u>, V. 1, Pergamon, 1967, pp. 95-120.

Struik, Dirk J., *Conic sections*, <u>Encyclopedia Americana</u>, 1976, V. 7, pp. 578-579.

Struik, Dirk J., *Differential geometry*, <u>Encyclopedia Americana</u>, 1976, V. 12, pp. 499-501.

Thébault, Victor, *Geodesics*, <u>Scripta Math.</u>, 21 (1955) 146-158.

Wylie, C. Ray, *Surface*, <u>Encyclopedia Americana</u>, 1976, V. 26, pp. 50-51.

5.3.2 ADVANCED

Banchoff, Thomas F., *Critical points and curvature for embedded polyhedral surfaces*, <u>Amer. Math. Monthly</u>, 77 (1970) 475-485.

> An explanation (with many diagrams) of the critical point theorem arising from Morse theory and the Gauss-Bonnet theorem in the special case of two dimensional surfaces. Some of the standard concepts involved are customarily defined using differential calculus. The author shows that not only can this be avoided, but that corresponding definitions and theorems can be given even for polyhedral surfaces.

Flanders, Harley, *Differential forms*, in S.S. Chern, <u>Studies in Global Geometry and Analysis</u>, M.A.A., 1967, pp. 57-95.

> An introductory survey of the use of differential forms in differential geometry. Includes a brief description of the relationship between differential forms and vector analysis, and a statement of Stokes's theorem and Gauss's theorem using differential forms.

Milnor, John, *A problem in cartography*, <u>Amer. Math. Monthly</u>, 76 (1969) 1101-1112.

> The author shows how to derive properties of a minimum distortion projection known to cartographers as the "azimuthal equidistant projection." In the process he is led to discuss various aspects of differential geometry. Winner of the Lester R. Ford award for expository writing.

Nitsche, Johannes C.C., *Plateau's problems and their modern ramifications*, <u>Amer. Math. Monthly</u>, 81 (1974) 945-968.

> Based on an address to the American Chemical Society, this paper describes some of Plateau's experiments on surface tension and the mathematical problems inspired by these experiments, typically, the search for surfaces of constant mean curvature subject to boundary constraints. The "soap-bubble" problem of finding a minimal surface (i.e., constant mean curvature zero) with a prescribed boundary is "Plateau's problem." Plateau's experimentally arrived at conjecture that for a simple closed curve in 3-space there is always a solution was proved in the 1930's. The paper describes other cases where there is no solution. Winner of the Lester R. Ford award for expository writing.

5.3.3 RESEARCH

Chern, S.S., *Curves and surfaces in Euclidean space*, in S.S. Chern, <u>Studies in Global Geometry and Analysis</u>, M.A.A., 1967, pp. 16-56.

> Aimed at the reader who already has some knowledge of elementary differential geometry of curves and surfaces. The emphasis is on global ("in the large") theorems about curves and surfaces in Euclidean 2-space and 3-space. Examples include the four vertex theorem, the isoperimetric inequality for plane curves, and the Gauss-Bonnet theorem. Winner of the Chauvenet Prize for expository writing in mathematics.

Chern, S.S., *Differential geometry*, <u>Encyclopaedia Britannica</u>, 15th ed., 1974, Macropaedia V. 7, pp. 1093-1099.

> This review of differential geometry includes such topics as the de Rham and Hodge theorems, the Gauss-Bonnet formula, characteristic classes, the Hirzebruch signature theorem, elliptic operators, etc.

Chern, S.S., *Differential geometry: its past and its future*, in Actes Cong. Inter. Math. (1970), V. 1, Gauthier-Villars, 1971, pp. 41-53.

Millman, R.S. and Stehney, Ann K., *The geometry of connections*, Amer. Math. Monthly, 80 (1973) 475-500.

> The authors develop the geometry of connections, starting with the definition of manifold and proceeding in precise textbook format. They close with the thesis that, even in the setting of Ehresmann's definition of connection, "...there is the dichotomy of Riemann--the concept of space (the principal fibre bundle) and the additional structure which yields the geometry (the connection), and that the geometric content of the massive structure of modern differential geometry is still apparent."

Nomizu, Katsumi, *Recent developments in the theory of connections and holonomy groups*, Advances in Math., 1 (1965) 1-49.

Viscensini, P., *Differential geometry in the nineteenth century*, Scientia, 107 (1972) 661-696.

> A thorough account with an extensive bibliography. The original article in French precedes this English translation, ibid., pp. 617-660.

RELATED REFERENCES

Cesari, Lamberto, *Recent results in surface area theory*, Amer. Math. Monthly, 66 (1959) 173-192. [4.2.2]

Cesari, Lamberto, *Surface area*, in S.S. Chern, Studies in Global Geometry and Analysis, M.A.A., 1967, pp. 123-146. [4.2.2]

Le Corbeiller, P., *The curvature of space*, Scientific American, 191 (November 1954) 80-86, 124; also in M. Kline, Mathematics in the Modern World, Freeman, 1968, pp. 128-133, 397. [1.2.0]

MacLane, Saunders, *Hamiltonian mechanics and geometry*, Amer. Math. Monthly, 77 (1970) 570-586. [7.1.2]

Munroe, M.E., *Bringing calculus up to date*, Amer. Math. Monthly, 65 (1958) 81-90. [4.3.2]

[5.3]

Serrin, James, *On the area of curved surfaces*, [4.3.2]
Amer. Math. Monthly, 68 (1961) 435-440.

Weyl, Hermann, *Relativity theory as a stimulus* [1.2.2]
in mathematical research, Proc. Amer. Phil.
Soc., 93 (1949) 535-541.

5.4 ALGEBRAIC GEOMETRY

5.4.2 ADVANCED

Bott, Raoul, *On the shape of a curve,* Advances in Math., 16 (1975) 144-159.

> The author's goal is to describe the recently proved (by Deligne) Weil conjectures. En route he gives examples to illustrate certain key aspects of algebraic geometry and topology.

Weil, André, *Algebraic geometry,* Encyclopedia Americana, 1976, V. 12, pp. 501-503.

5.4.3 RESEARCH

Algebraic geometry, Encyclopaedia Britannica, 15th ed., 1974, Macropaedia V. 7, pp. 1070-1076.

> This seven page review of algebraic geometry is decidedly at the graduate level: consider such topics as Weil's generalization of a projective variety, the Chow ring, Grothendieck's generalization of an algebraic variety, and a "borrowing" of Chern classes from differential geometry.

Dieudonné, Jean A., *Algebraic geometry,* Advances in Math., 3 (1969) 233-321.

> Winner of the LeRoy P. Steele prize for distinguished exposition of outstanding research.

Emerson, John D., *Simple points of an affine algebraic variety,* Amer. Math. Monthly, 82 (1975) 132-147.

> The author reviews various aspects of algebraic geometry en route to his objective, which is to present a proof of a 1947 result of Zariski concerning simple points on an affine algebraic variety, to wit that an abstract definition of simple point coincides "in a rather general setting" with the classical notion.

Griffiths, Phillip A., *Periods of integrals on algebraic manifolds,* Bull. Amer. Math. Soc., 76 (1970) 228-296.

> Winner of the LeRoy P. Steele prize for distinguished exposition of outstanding research.

Kleiman, S.L. and Laksov, Dan, *Schubert calculus,* Amer. Math. Monthly, 79 (1972) 1061-1082.

> "How many lines in 3-space, in general, intersect four given lines?" The authors use this type of enumerative problem (which

was considered by H. Schubert a century ago) to motivate and illustrate a discussion of a portion of algebraic geometry. After a few pages cohomology theory sets in, but in the interim various basic notions involving determinants, projective space, Plucker coordinates, and Grassman manifolds of d-planes in n-space are described in the context of the enumerative problem.

Lefschetz, Solomon, *A page of mathematical autobiography*, Bull. Amer. Math. Soc., 74 (1968) 854-879.

Winner of the LeRoy P. Steele prize for distinguished exposition of outstanding research.

RELATED REFERENCES

Cassels, J.W.S., *Diophantine equations with special reference to elliptic curves*, J. London Math. Soc., 41 (1966) 193-291. [3.3.3]

Dieudonné, Jean A., *The historical development of algebraic geometry*, Amer. Math. Monthly, 79 (1972) 827-866. [1.2.3]

Lefschetz, Solomon, *The early development of algebraic geometry*, Amer. Math. Monthly, 76 (1969) 451-460. [1.2.3]

Novikov, S.P., *The main trends of algebraic topology and algebraic geometry*, Russian Math. Surveys, 19:6 (1964) 67-74. [5.7.3]

Reyes, Gonzalo E., *From sheaves to logic*, in A. Daigneault, Studies in Algebraic Logic, M.A.A., 1974, pp. 143-204. [2.6.3]

Tate, J., *The 1974 Fields Medals (I): An algebraic geometer*, Science, 186 (1974) 39-40. [1.3.2]

van der Waerden, B.L., *The foundation of algebraic geometry from Severi to André Weil*, Arch. Hist. Exact Sci., 7 (1971) 171-180. [1.2.3]

5.5 GENERAL TOPOLOGY

5.5.0 GENERAL

Barr, Stephen, Experiments in Topology, Thomas Crowell, 1964.

Bing, R.H., *Point-set topology*, in The Mathematical Sciences--A Collection of Essays for COSRIMS, M.I.T., 1969, pp. 209-216.

Courant, Richard and Robbins, Herbert, *Topology*, in R. Courant and H. Robbins, What is Mathematics, Oxford U. Pr., 1941, pp. 235-271; also in J.R. Newman, The World of Mathematics, V. 1, Simon and Schuster, 1956, pp. 581-599.

> A survey of popular topics in topology--Euler's formula for polyhedra, rubber-sheet geometry, Jordan curve theorem, four color problem, classification of surfaces.

Franklin, Philip, *What is topology?*, Phil. Sci., 2 (1935) 39-47.

Lietzmann, W., Visual Topology, American Elsevier, 1965.

> Intriguing challenging approach using many drawings and illustrations to introduce concepts of elementary topology.

Tucker, Albert W. and Bailey, Herbert S., Jr., *Topology*, Scientific American, 182 (January 1950) 18-24, 64; also in M. Kline, Mathematics in the Modern World, Freeman, 1968, pp. 134-140, 398.

> In their excellent book First Concepts of Topology, W.G. Chinn and N.E. Steenrod inveigh against parlor tricks which "are a caricature of topology." If having been warned against sin you want to see for yourself, you will find in this popularization a detailed--and illustrated--recipe for turning an inner tube (with a valve hole) inside out. There is also a description of the Kline bottle and its interpretation as a pair of Möbius bands glued together, as well as discussions of various other topological notions.

5.5.1 ELEMENTARY

Aleksandrov, P.S., *Topology*, in A.D. Aleksandrov, et al., Mathematics--Its Content, Methods and Meaning, V. 3,

M.I.T., 1963, pp. 193-225.

> An elementary account of topology starting with the diagrams of a torus, a Möbius strip, and a Klein bottle. The notion of index of a singular point is used to indicate a relationship between the qualitative theory of differential equations and topology, and also to demonstrate Brouwer's fixed point theorem for a disk. The Euler characteristic of compact oriented surfaces is computed in a few sentences.

Chinn, William G. and Steenrod, Norman E., <u>First Concepts of Topology</u>, New Math. Libr., No. 18, Random House, 1966; M.A.A., 1975.

> High powered material for high school students, but for the person with some training in mathematics it is a superb introduction to topology. Many helpful diagrams illustrate the text.

Flegg, Graham, <u>From Geometry to Topology</u>, Crane Russak, 1974.

Hilbert, David and Cohn-Vossen, Stephan, <u>Geometry and the Imagination</u>, Chelsea, 1952.

> An interesting expository account of basic geometric and topological notions.

Kline, J.R., *What is the Jordan curve theorem?*, <u>Amer. Math. Monthly</u>, 49 (1942) 281-286.

Menger, Karl, *What is dimension?*, <u>Amer. Math. Monthly</u>, 50 (1943) 2-7.

> Discusses the desiderata of dimension theory.

Moore, John C., *Topology*, <u>McGraw-Hill Encyclopedia of Science and Technology</u>, 1960, V. 13, pp. 679-683.

Phillips, Anthony, *Topology*, <u>Encyclopedia Americana</u>, 1976, V. 26, pp. 850-854.

Shinbrot, Marvin, *Fixed-point theorems*, <u>Scientific American</u>, 214 (January 1966) 105-110, 136; also in M. Kline, <u>Mathematics in the Modern World</u>, Freeman, 1968, pp. 145-150, 398.

> This popular article presents intuitive proofs of Brouwer's two-dimensional fixed point theorem and its one-dimensional analog. There is also a brief discussion of the use of infinite-dimensional fixed point theory to solve the differential equations of dynamics, but because of the lack of detail this portion is difficult to follow.

Wilder, Raymond L., *Topology--its nature and significance*, <u>Math. Teacher</u>, 55 (1962) 462-475.

5.5.2 ADVANCED

Aleksandrov, P.S., *Poincaré and topology*, <u>Russian Math. Surveys</u>, 27:1 (1972) 157-168.

"He created it." Transcription of speech given in The Hague at the International Congress of Mathematicians in 1954. Very easy reading.

Aleksandrov, P.S., *Some results in the theory of topological spaces, obtained within the last 25 years*, <u>Russian Math. Surveys</u>, 15:2 (1960) 23-84.

Bing, R.H., *Challenging conjectures*, <u>Amer. Math. Monthly</u>, 74 (1967) Suppl. pp. 56-64.

Between exhortations to mathematics teachers to stay professionally alive, the author sketches the statements of five famous unsolved problems: the continuum hypothesis (despite Cohen's results), the normal Moore space conjecture, the fixed point conjecture for cellular plane sets, the Poincaré conjecture, and the annulus conjecture.

Bing, R.H., *Elementary point set topology*, <u>Amer. Math. Monthly</u>, 67 (1960) Suppl. pp. 1-58.

This article is an unusual mixture of popularization, precision, and intuitive description. Although the first few sections are directed at the rank beginner ("a triangle and a 'wiggly' simple closed curve are alike in some respects"), the overall article is probably best suited to the person who knows a little point set topology. Includes exercises throughout and a good bibliography.

Bing, R.H., *General topology*, <u>Encyclopaedia Britannica</u>, 15th ed., 1974, Macropaedia V. 18, pp. 509-514.

Includes a section on problems of current interest.

Bing, R.H., *Spheres in E-3*, <u>Amer. Math. Monthly</u>, 71 (1964) 353-364.

Many illustrations make this a good source for topological intuition. Winner of the Lester R. Ford award for expository writing.

Brown, Robert F., *Elementary consequences of the noncontractibility of the circle*, <u>Amer. Math. Monthly</u>, 81 (1974) 247-252.

The author's thesis is that a good undergraduate point set topology course provides all the background necessary for an elementary (no algebraic topology) proof of the circle's noncontractibility, as well as numerous consequences--such as the Jordan curve theorem and the Brouwer fixed point theorem. He demonstrates parts of this thesis and cites references for other parts. (One could wish that he had included the cited proofs in his paper.)

McAllister, B.L., *Cyclic elements in topology: a history*, <u>Amer. Math. Monthly</u>, 73 (1966) 337-350.

> A Peano continuum is a metric space which is the continuous image of a closed interval. The reason for abandoning the original name "continuous curve" was that Peano's space-filling curve showed that the unit square satisfies the above definition. Cyclic elements were introduced by G.T. Whyburn in 1926 as a device for analyzing and classifying continua. The present paper outlines the subsequent work based on this notion. The non-specialist may be surprised by the great number of prominent mathematicians involved in this history.

Rudin, Mary Ellen, *Souslin's conjecture*, <u>Amer. Math. Monthly</u>, 76 (1969) 1113-1119.

> Souslin's 1920 conjecture (a proposed topological characterization of the real line) is related to mathematical logic, set theory, and topology. The author refers to "the spell of Souslin's conjecture," and that would be a descriptive characterization of her paper.

Steen, Lynn Arthur, *Conjectures and counterexamples in metrization theory*, <u>Amer. Math. Monthly</u>, 79 (1972) 113-132.

> "...as metrization research shifts from topology to logic, we survey in this paper the chief topological milestones [related to metrization] of the last half century." The survey emphasizes the role of the Moore space conjecture. It is written tersely, but provided with diagrams to summarize the interrelationships among the many definitions and theorems that are given. Fifteen exotic topologies are described which serve as the counterexamples of the title and are keyed to the survey. Winner of the Lester R. Ford award for expository writing.

Tucker, Albert W., *Some topological properties of disk and sphere*, in <u>Proc. First Canad. Math. Cong.</u>, U. Toronto Pr., 1946, pp. 285-309.

Whyburn, Gordon T., *What is a curve?*, <u>Amer. Math. Monthly</u>, 49 (1942) 493-497; also in P.J. Hilton, <u>Studies in Modern Topology</u>, M.A.A., 1968, pp. 23-38.

5.5.3 RESEARCH

Aleksandrov, P.S., *On some basic directions in general topology*, <u>Russian Math. Surveys</u>, 19:6 (1964) 1-40.

Bing, R.H., *The elusive fixed point property*, Amer. Math. Monthly, 76 (1969) 119-132.

> "The paper is intended primarily as an expository article...about fixed points... . [A] graduate student with a beginning course in geometric topology [should] be able to dig through most of the material."

Whyburn, Gordon T., *On the structure of continua*, Bull. Amer. Math. Soc., 42 (1936) 49-73.

> Winner of the Chauvenet Prize for expository writing in mathematics.

RELATED REFERENCES

Buck, R.C., *Topology and analysis*, Math. Magazine, 40 (1967) 71-74. [4.2.2]

Dauben, Joseph W., *Denumerability and dimension: the origins of Georg Cantor's theory of sets*, Rete, 2 (1974) 105-133. [1.2.2]

Dauben, Joseph W., *The invariance of dimension: problems in the early development of set theory and topology*, Historia Math., 2 (1975) 273-288. [1.2.2]

Dauben, Joseph W., *The trigonometric background to Georg Cantor's theory of sets*, Arch. Hist. Exact Sci., 7 (1971) 181-216. [1.2.2]

Hewitt, Edwin, *The role of compactness in analysis*, Amer. Math. Monthly, 67 (1960) 499-516. [4.3.2]

Hildebrandt, T.H., *The Borel theorem and its generalizations*, Bull. Amer. Math. Soc., 32 (1926) 423-474. [4.3.3]

Manheim, Jerome H., The Genesis of Point Set Topology, Pergamon, 1964. [1.2.2]

McShane, E.J., *Partial orderings and Moore-Smith limits*, Amer. Math. Monthly, 59 (1952) 1-11. [4.3.2]

Whyburn, Gordon T., *Developments in topological analysis*, Fund. Math., 50 (1962) 305-318. [4.4.2]

Whyburn, Gordon T., *Topological analysis*, [4.3.2]
Bull. Amer. Math. Soc., 62 (1956) 204-218.

Wilder, Raymond L., *The origin and growth of* [1.5.1]
mathematical concepts, Bull. Amer. Math. Soc.,
59 (1963) 423-448.

5.6 DIFFERENTIAL TOPOLOGY

5.6.1 ELEMENTARY

Phillips, Anthony, *Turning a surface inside out*, <u>Scientific American</u>, 214 (May 1966) 112-120, 148.
  Somewhat advanced ideas of differential topology.

Stewart, Ian, *The seven elementary catastrophes*, <u>New Scientist</u> (29 November 1975) 447-454.

Zeeman, E. Christopher, *The geometry of catastrophe*, <u>Times Lit. Suppl.</u>, (10 December 1971) 1556-1557.

5.6.2 ADVANCED

Bing, R.H., *Some aspects of the topology of 3-manifolds related to the Poincaré conjecture*, in T.L. Saaty, <u>Lectures on Modern Mathematics</u>, V. 2, Wiley, 1964, pp. 93-128.

Callahan, James, *Singularities and plane maps*, <u>Amer. Math. Monthly</u>, 81 (1974) 211-240.
  "What structure does [a differentiable map of the plane into itself] have around a point where the Jacobian is zero?" The author addresses this and related questions for pairs of dimensions other than (2,2). He thus discusses (and illuminates with copious diagrams) work of Morse, Whitney, Thom, and Mather. He mentions the relevance of singularities to the Thom-Zeeman theory of catastrophes and its biological applications, but he does not pursue this aspect (references to germs, swallowtails, and butterflies notwithstanding). Winner of the Lester R. Ford award for expository writing.

Morse, Marston, *What is analysis in the large?*, <u>Amer. Math. Monthly</u>, 49 (1942) 358-364; also in S.S. Chern, <u>Studies in Global Geometry and Analysis</u>, M.A.A., 1967, pp. 5-15.
  This paper succeeds admirably in using a diverse series of examples to get across the notion of "in the large." There is a brief discussion of Poincaré's early work on differential equations in this context. The paper closes with some elementary (but surprising) geometrical consequences of Morse's critical point theorem.

Seebach, J. Arthur, Jr., Seebach, Linda A. and Steen, Lynn Arthur, *What is a sheaf?*, Amer. Math. Monthly, 77 (1970) 681-703.

The authors' aim is to present an exposition of sheaf theory without the use of sophisticated algebraic topology. Their approach is to give examples of sheaves of germs of holomorphic functions (of several complex variables), sheaves of local rings, and sheaves of differential forms as motivation before giving the general definition. They conclude: "Thus sheaves transform a complicated property of functions, such as analyticity, into the simpler one of continuity, for the topology of the sheaf is chosen precisely so that a continuous section on U...corresponds to one of the specialized (e.g., analytic, differentiable) functions of F(U)."

Smale, Stephen, *What is global analysis?*, Amer. Math. Monthly, 76 (1969) 4-9.

"My definition of global analysis is simply the study of differential equations, both ordinary and partial, on manifolds and vector space bundles. Thus one might consider global analysis as differential equations from a global, or topological, point of view." After showing how to set up the topological space of differential equations on a manifold, Smale describes a theorem of Peixoto on structural stability.

5.6.3 RESEARCH

Cohen, Maurice, *Foliations of 3-manifolds*, Amer. Math. Monthly, 81 (1974) 462-473.

This paper describes the progress made on foliations of manifolds during the quarter-century 1944-69, with some emphasis on the starting point: Reeb's foliation of the 3-sphere. The subject matter is profusely illustrated, making the material more accessible. Discusses the historical origin of the subject in differential equations and the use of the Poincaré-Bendixson theorem in the development of the theory.

Lashof, Richard, *The tangent bundle of a topological manifold*, Amer. Math. Monthly, 79 (1972) 1090-1096.

Based on Milnor's notion of microbundle, this article describes how to generalize the notion of tangent bundle to topological manifolds, i.e., to manifolds where the differential manifold definition of tangent vector breaks down because there is no differentiability. While the article would be best read by those who are already familiar with the differential case, it contains a condensed review of the necessary material concerning differential manifolds.

Lawson, H. Blaine, *Foliations*, <u>Bull. Amer. Math. Soc.</u>, 80 (1974) 369-418.

   Winner of the LeRoy P. Steele prize for distinguished exposition of outstanding research.

Milnor, John, *A survey of cobordism theory*, <u>L'Enseignement Math.</u>, 8 (1962) 16-23.

   A survey of work on the widely-used fundamental notion of cobordism which Rene Thom introduced into the study of differentiable manifolds.

Milnor, John, *Differential topology*, in T.L. Saaty, <u>Lectures on Modern Mathematics</u>, V. 2, Wiley, 1964, pp. 165-183.

Poénaru, Valentin, *On the geometry of differentiable manifolds*, in P.J. Hilton, <u>Studies in Modern Topology</u>, M.A.A., 1968, pp. 165-207.

Smale, Stephen, *A survey of some recent developments in differential topology*, <u>Bull. Amer. Math. Soc.</u>, 69 (1963) 131-145.

Smale, Stephen, *Differentiable dynamical systems*, <u>Bull. Amer. Math. Soc.</u>, 73 (1967) 747-817.

RELATED REFERENCES

Gleason, Andrew M., *Evolution of an active mathematical theory*, <u>Science</u>, 145 (1964) 451-457; also appeared as *The evolution of differential topology*, in <u>The Mathematical Sciences--A Collection of Essays for COSRIMS</u>, M.I.T., 1969, pp. 176-189. [1.2.1]

Hilton, Peter J., *Introduction: modern topology*, in P.J. Hilton, <u>Studies in Modern Topology</u>, M.A.A., 1968, pp. 1-22. [5.7.3]

Mostow, George D. and Thom, René F., *Topological groups and differential topology*, <u>Encyclopaedia Britannica</u>, 15th ed., 1974, Macropaedia V. 18, pp. 489-504. [5.8.3]

Panati, Charles, *Catastrophe theory*, <u>Newsweek</u> (January 19, 1976) 54-55. [7.3.0]

Thom, René F., *Topological models in biology*, <u>Topology</u>, 8 (1969) 313-335. [7.2.2]

Zeeman, E. Christopher, *Catastrophe theory*, [7.2.1]
Scientific American, 234 (April 1976) 65-83,
138.

Zeeman, E. Christopher, *Levels of structure* [7.3.2]
in catastrophe theory illustrated by applications in the social and biological sciences,
in Proc. Inter. Cong. Math. (1974), V. 2,
Canad. Cong. Math., 1975, pp. 533-546.

5.7 ALGEBRAIC TOPOLOGY

5.7.2 ADVANCED

Algebraic topology, Encyclopaedia Britannica, 15th ed., 1974, Macropaedia V. 18, pp. 504-509.

> A concise presentation of material usually found in a graduate level course (homotopy theory, homology and cohomology, spectral sequences); the article concludes with a brief resumé of recent advances.

Atiyah, M.F., *The role of algebraic topology in mathematics*, J. London Math. Soc., 41 (1966) 63-69.

> An overview of the interaction of algebraic topology with such topics as Riemann surface theory, homological algebra, algebraic geometry, elliptic differential operators.

Hilton, Peter J., *Topology in the high school*, Educ. Studies Math., 3 (1970-71) 436-453.

> Good intuitive interpretation of theorems about the fundamental group; harder than he suggests.

5.7.3 RESEARCH

Bott, Raoul, *The periodicity theorem for the classical groups and some of its applications*, Advances in Math., 4 (1970) 353-411.

> A graduate-level exposition which traces the development of the Bott periodicity theorems, and leads to a description of K-theory.

Curtis, Edward B., *Simplicial homotopy theory*, Advances in Math., 6 (1971) 107-209.

> Winner of the LeRoy P. Steele prize for distinguished exposition of outstanding research.

Dyer, Eldon, *The functors of algebraic topology*, in P.J. Hilton, Studies in Modern Topology, M.A.A., 1968, pp. 134-164.

Eilenberg, Samuel, *Algebraic topology*, in T.L. Saaty, Lectures on Modern Mathematics, V. 1, Wiley, 1963, pp. 98-114.

> The author uses the problem of ascertaining the number of linearly independent non-vanishing vector fields on an n-sphere as a vehicle

to introduce some of the machinery of algebraic topology. Except for the first several pages, the article is intended strictly for mathematicians.

Gugenheim, V.K.A.M., *Semisimplicial homotopy theory*, in P.J. Hilton, Studies in Modern Topology, M.A.A., 1968, pp. 99-133.

Hilton, Peter J., *Introduction: modern topology*, in P.J. Hilton, Studies in Modern Topology, M.A.A., 1968, pp. 1-22.

This article goes beyond its role as an introduction and presents a summary of major advances in topology since World War II. The author succeeds in giving brief but clear statements of many of the key results of this period, as well as organizing much of the material in a coherent scheme. Some knowledge of topology is presupposed.

Hilton, Peter J., *Localization in topology*, Amer. Math. Monthly, 82 (1975) 113-131.

The author begins by recounting the localization theory of abelian groups, and then shows how this theory applies both as a model and as a tool in developing a corresponding theory in topology. New results from the theory include certain non-cancellation phenomena (e.g., knowing $X \times S^3$ up to homotopy does not prescribe X).

Novikov, S.P., *The main trends of algebraic topology and algebraic geometry*, Russian Math. Surveys, 19:6 (1964) 67-74.

Schultz, Reinhard, *Some recent results on topological manifolds*, Amer. Math. Monthly, 78 (1971) 941-952.

The author summarizes basic results concerning topological manifolds and piecewise linear manifolds, including brief discussions of the triangulation conjecture, the Hauptvermutung, and handle body theory--as well as older results going back to the Jordan curve theorem.

Seifert, Herbert and Threlfall, William, *Old and new results on knots*, Canad. J. Math., 2 (1950) 1-15.

Extract from a review by R.H. Fox: "A carefully reasoned development of the central part of knot theory (the group knot, the Alexander polynomial, the homology groups and linking invariants of the cyclic coverings) together with explanation of new results...this survey should furnish an excellent guide to the nonspecialist. The combinatorial point of view prevails... ."

Steenrod, Norman E., *Cohomology operations and obstructions to extending continuous functions*, Advances in Math., 8 (1972) 371-416.

RELATED REFERENCES

Baylis, E.R., *Knots--a practical application of group theory*, Math. Gazette, 57 (1973) 311-320. [3.5.2]

Bott, Raoul, *On the shape of a curve*, Advances in Math., 16 (1975) 144-159. [5.4.2]

Elder, Barbara, *Paths and knots as geometric groups*, Pentagon, 28 (1968) 3-15. [3.5.2]

Lakatos, Imre, *Proofs and refutations*, Brit. J. Phil. Science, 14 (1963-64) 1-25, 120-139, 221-245, 296-342. [1.5.2]

Lefschetz, Solomon, *A page of mathematical autobiography*, Bull. Amer. Math. Soc., 74 (1968) 854-879. [5.4.3]

Redheffer, Raymond, *The homotopy theorems of function theory*, Amer. Math. Monthly, 76 (1969) 778-787. [4.4.2]

5.8 Topological Groups

5.8.2 ADVANCED

Kaplansky, Irving, *Topological rings*, Bull. Amer. Math. Soc., 54 (1948) 809-826.

Mackey, George W., *Group representations and analysis*, Rice Univ. Studies, 49:4 (1963) 13-27.
 Entirely self contained survey article; no references.

Montgomery, Deane, *What is a topological group?*, Amer. Math. Monthly, 52 (1945) 302-307.

5.8.3 RESEARCH

Belinfante, Johan G.F. and Kolman, Bernard, *An introduction to Lie groups and Lie algebras, with applications, I-III*, SIAM Review, 8 (1966) 11-46; 10 (1968) 160-195; 11 (1969) 510-543; revised and reprinted as A Survey of Lie Groups and Lie Algebras, SIAM, 1972.

Freudenthal, Hans, *Lie groups in the foundations of geometry*, Advances in Math., 1 (1965) 145-190.

Mackey, George W., *Infinite-dimensional group representations*, Bull. Amer. Math. Soc., 69 (1963) 628-686.
 Long survey with excellent bibliography.

Mayer, W. and Thomas, T.Y., *Foundations of the theory of Lie groups*, Annals of Math., 36 (1935) 770-822.
 First comprehensive global treatment--i.e., as groups. Few references.

Mostow, George D. and Thom, René F., *Topological groups and differential topology*, Encyclopaedia Britannica, 15th ed., 1974, Macropaedia V. 18, pp. 489-504.
 The first section, on topological methods in analysis, is a thorough treatment of Lie groups and related concepts, emphasizing "the three-fold structure of continuous groups--algebraic, topological, and analytic... ." The second section is a concise presentation of the basic concepts, problems and methods in differential topology.

Salam, Abdus, *Theory of groups and the symmetry physicist*, J. London Math. Soc., 41 (1966) 49-62.
  Survey of non-compact groups and particle physics. Exhibits some hidden symmetries in the elementary particles and outlines current work involving SU(2) and more complicated groups.

RELATED REFERENCES

Bolker, Ethan D., *The spinor spanner*, Amer. Math. Monthly, 80 (1973) 977-984.   [7.1.2]

Curtis, Charles W., *The classical groups as a source of algebraic problems*, Amer. Math. Monthly, 74 (1967) Suppl. pp. 80-91.   [3.5.3]

Hawkins, Thomas, *Hypercomplex numbers, Lie groups and the creation of group representation theory*, Arch. Hist. Exact Sci., 8 (1972) 243-287.   [1.2.3]

Hoffman, William C., *Visual illusions of angle as an application of Lie transformation groups*, SIAM Review, 13 (1971) 169-184.   [7.2.2]

Mackey, George W., *Functions on locally compact groups*, Bull. Amer. Math. Soc., 56 (1950) 385-412.   [4.7.3]

6.1 PROBABILITY

6.1.0 GENERAL

Carnap, Rudolf, *What is probability?*, Scientific American, 189 (September 1953) 128-138, 170; also in D. Messick, Mathematical Thinking in Behavioral Sciences, Freeman, 1968, pp. 14-17, 223.

> Argues that there are two kinds, statistical and inductive.

Cohen, John, *Subjective probability*, Scientific American, 197 (November 1957) 128-138, 184; also in D. Messick, Mathematical Thinking in Behavioral Sciences, Freeman, 1968, pp. 18-22, 223.

> Explores how personal judgments affect the way people evaluate the probabilities of events. Experiments determine how closely these subjective values conform to actual probability laws.

Huff, Darrell, How to Take a Chance, W.W. Norton, 1959.

> A lively and imaginative book, written at a popular level of poker hands, coin tossing, courtroom evidence, etc. It has very little mathematics, but it might help non-probabilists to think in probabilistic terms.

Kac, Mark, *Probability*, Scientific American, 211 (September 1964) 92-108, 269; also in M. Kline, Mathematics in the Modern World, Freeman, 1968, pp. 165-174, 398-399; in D. Messick, Mathematical Thinking in Behavioral Sciences, Freeman, 1968, pp. 23-32, 223-224; and in The Mathematical Sciences--A Collection of Essays for COSRIMS, M.I.T., 1969, pp. 232-251.

> Quoting Laplace with loving approval, "it is remarkable that a science which began with the consideration of games of chance should have become the most important object of human knowledge," the author discusses various aspects of probability, e.g., normal curve, Brownian motion, Buffon needle problem, and Clausian heat death.

Leibowitz, Martin A., *Queues*, Scientific American, 219 (August 1968) 96-103, 124.

> Mathematical analysis of queues.

Molina, Edward C., *Bayes' theorem--an expository presentation*, Annals of Math. Stat., 2 (1931) 23-37.

Nagel, Ernest, *The meaning of probability*, <u>J. Amer. Stat. Assoc.</u>, 31 (1936) 10-30; also in J.R. Newman, <u>The World of Mathematics</u>, V. 2, Simon and Schuster, 1956, pp. 1398-1414.

> Detailed, occasionally tedious exploration of the meaning of probability as related to its context. Nagel offers several different interpretations by considering (1) everyday discourse, (2) applied statistics and measurements, (3) physical and biological theories, (4) the comparison of various other theories, and (5) the calculus of probability. Finally, he concludes by proposing that the frequency interpretation is best.

Ore, Oystein, *Pascal and the invention of probability theory*, <u>Amer. Math. Monthly</u>, 67 (1960) 409-419.

Page, David A., *Probability*, <u>NCTM Twenty-Fourth Yearbook</u>, 1959, 229-271.

Schlaifer, Robert, *The meaning of probability*, in B. Lieberman, <u>Contemporary Problems in Statistics</u>, Oxford U. Pr., 1971, pp. 236-249.

Steen, Lynn Arthur, *Order from chaos*, <u>Science News</u>, 107 (1975) 292-293.

Udell, Dan E., *Self-similar models for erratic chance processes*, <u>IBM Research Reports</u>, 2:2 (1966) 1-4.

von Mises, Richard, <u>Probability, Statistics and Truth</u>, Macmillan, 1957.

Weaver, Warren, *Probability*, <u>Scientific American</u>, 183 (October 1950) 44-47, 64; also in M. Kline, <u>Mathematics in the Modern World</u>, Freeman, 1968, pp. 161-164, 398.

> Distinguishes probability from chance. Thesis: probability plays the role in science that faith plays in human activity. The only time we can be really sure is when we are dealing with matters of faith, not those of science--because those of science are based on chance.

Youden, William J., *How mathematics appraises risks and gambles*, in T.L. Saaty and F.J. Weyl, <u>The Spirit and Uses of the Mathematical Sciences</u>, McGraw-Hill, 1969, pp. 167-187.

6.1.1 ELEMENTARY

Ayer, A.J., *Chance*, <u>Scientific American</u>, 213 (October 1965) 44-54, 126-127; also in M. Kline, <u>Mathematics in the Modern</u>

[6.1]

World, Freeman, 1968, pp. 151-160, 398; and in D. Messick, *Mathematical Thinking in Behavioral Sciences*, Freeman, 1968, pp. 4-13, 223.

> Moderately difficult article which classifies chance into a priori probability, statistical judgment and judgments of credibility. Includes interesting illustrations.

Doob, J.L., *What is a stochastic process?*, Amer. Math. Monthly, 49 (1942) 648-653.

Feller, William, *Probability*, McGraw-Hill Encyclopedia of Science and Technology, 1960, V. 10, pp. 624-630.

Freudenthal, Hans, *Models in applied probability*, Synthese, 12 (1960) 202-212.

> The seemingly indispensable urn model can be replaced in applied probability by the model of the "complete strategist" playing a game against another "complete strategist."

Grenander, Ulf, *Computational probability and statistics*, SIAM Review, 15 (1973) 134-192.

Hammersley, J.M., *Some speculations of a sense of nicely calculated chances*, SIAM Review, 16 (1974) 237-255.

Hersh, Reuben and Griego, Richard J., *Brownian motion and potential theory*, Scientific American, 220 (March 1969) 66-74, 148.

> "The discovery that these two apparently unrelated branches of physics are in some sense mathematically equivalent has led to a new subject known as probabilistic potential theory."

Kolmogorov, A.N., *The theory of probability*, in A.D. Aleksandrov, et al., Mathematics--Its Content, Methods and Meaning, V. 2, M.I.T., 1963, pp. 229-264.

Munroe, M.E., *Mathematical theory of probability*, Encyclopedia Americana, 1976, V. 22, pp. 622-625.

Struik, Dirk J., *On the foundations of the theory of probabilities*, Phil. Sci., 1 (1934) 50-70.

Theory of probability, Encyclopaedia Britannica, 15th ed., 1974, Macropaedia V. 14, pp. 1104-1115.

> Survey of basic concepts and standard theorems; includes sections on various methods of reasoning in this discipline, advanced probability, and stochastic processes.

6.1.2 ADVANCED

Bhat, U. Narayan, *Sixty years of queueing theory*, Management Science, 15:6 (1969) B280-B294.

Boudreau, P.E., Griffin, J.S., Jr. and Kac, Mark, *An elementary queueing problem*, Amer. Math. Monthly, 69 (1962) 713-724.

> The authors' aim is to illustrate their point that probability theory contains many interesting examples of "mathematics in action." They use a Markov chain, some complex analysis, and a Tauberian theorem to solve their queueing problem.

Doob, J.L., *What is a martingale?*, Amer. Math. Monthly, 78 (1971) 451-463.

> "Martingale theory illustrates the history of mathematical probability: the basic definitions are inspired by crude notions of gambling, but the theory has become a sophisticated tool of modern abstract mathematics, drawing from and contributing to other fields...The following account...is designed to give a feeling for the subject with a minimum of technicality."

Feller, William, *Chance processes and fluctuations*, in E.F. Beckenbach, Modern Mathematics for the Engineer, 2nd Ser., McGraw-Hill, 1961, pp. 167-181.

> A study of two "methodologically interesting" problems. Feller's approach is to give heuristic demonstrations starting with a random walk problem, as in his derivation of the normal distribution as the fundamental solution of the heat equation. He uses the same approach to discuss queueing problems, and closes by inveighing against the term "statistical equilibrium."

Feller, William, *The fundamental limit theorems in probability*, Bull. Amer. Math. Soc., 51 (1945) 800-832.

> Entirely self-contained; discusses the mathematical content of the central limit theorem and the Khinchin-Kolmogorov Law of the iterated log. Many references.

Gridgeman, N.T., *Geometric probability and the number pi*, Scripta Math., 25 (1960) 183-195.

> Discussion of the Buffon needle method and its extensions, together with debunking of suspect data of Lazzerini and Fox of trials of the method.

Halmos, Paul R., *The foundations of probability*, Amer. Math. Monthly, 51 (1944) 493-510.

> Winner of the Chauvenet Prize for expository writing in mathematics.

Jeffreys, Harold, *The present position in probability theory*, Brit. J. Phil. Sci., 5 (1955) 275-289.

Kac, Mark, *Random walk and the theory of Brownian motion*, Amer. Math. Monthly, 54 (1947) 369-391.

> Extensive references; reaches up to then-current research. Winner of the Chauvenet Prize for expository writing in mathematics.

Kendall, David G., *Branching processes since 1873*, J. London Math. Soc., 41 (1966) 385-406.

> "The term 'branching process' appears to have been coined by A.N. Kolmogorov and N.A. Dmitriev in 1947 to describe the stochastic processes which arise when the theory of probability is introduced into population mathematics." Making effective use of references to the extinction of peerages, the author traces the history of the subject before 1947 and the development since.

Kendall, David G., *The genealogy of genealogy: branching processes before (and after) 1873*, Bull. London Math. Soc., 7 (1975) 225-253.

Koopman, Bernard Osgood, *The axioms and algebra of intuitive probability*, Annals of Math., 41 (1940) 269-292.

Kuller, Robert G., *Coin tossing, probability and the Weierstrass approximation theorem*, Math. Magazine, 37 (1964) 262-265.

> Winner of the Lester R. Ford award for expository writing.

Langford, Eric, *A problem in geometric probability*, Math. Magazine, 43 (1970) 237-244.

> Winner of the Lester R. Ford award for expository writing.

Neuts, Marcel F., *Are many 1-1 functions on the positive integers onto?*, Math. Magazine, 41 (1968) 103-109.

> Winner of the Lester R. Ford award for expository writing.

Robbins, Herbert, *Optimal stopping*, Amer. Math. Monthly, 77 (1970) 333-343.

> "The theory of probability began with efforts to calculate the odds in games of chance. In this context, optimal stopping problems concern the effect on a gambler's fortune of various possible systems for deciding when to stop playing a sequence of games. Such problems are of interest in statistics, where the experimenter must constantly ask whether the increase in information contained in further data will outweigh the cost of collecting it."

Robbins, Herbert, *The theory of probability*, NCTM Twenty-Third Yearbook, 1957, 336-371.

> A scholarly and somewhat technical overview.

Ulam, Stanislaw M., *Monte Carlo calculations in problems of mathematical physics*, in E.F. Beckenbach, Modern Mathematics for the Engineer, 2nd Ser., McGraw-Hill, 1961, pp. 261-281.

Consider $n^2 + 1$ integers, arranged in all possible permutations. Each arrangement is known to have a monotone subsequence of length at least n, but the average length (of the longest monotone subsequence) will be approximately 1.7n. At least, so it has been conjectured on the basis of Monte Carlo experiments. By this example (and an estimation of an integral in 10-dimensional space) the author gives the uninitiated a feel for the Monte Carlo method.

6.1.3 RESEARCH

Loève, Michel, *On stochastic processes*, in T.L. Saaty, Lectures on Modern Mathematics, V. 3, Wiley, 1965, pp. 245-276.

"Three propositions play a fundamental role in probability theory and its applications... ." The three are (roughly) a theorem of Doob on martingale convergence, the von Neumann and Birkhoff ergodic theorems, and infinite decomposability theorems of Khintchine, Lévy, and Gnedenko. The paper elaborates on this assertion.

Moerbeke, P.V., *Optimal stopping and free boundary problems*, Rocky Mountain J. Math., 4 (1974) 539-578.

A leisurely, well-written account.

RELATED REFERENCES

Barr, Donald R., *When will the next record rain-* [7.1.2]
fall occur?, Math. Magazine, 45 (1972) 15-19.

Billingsley, Patrick, *On the central limit* [3.3.2]
theorem for the prime divisor function,
Amer. Math. Monthly, 76 (1969) 132-139.

Billingsley, Patrick, *Prime numbers and* [3.3.2]
Brownian motion, Amer. Math. Monthly, 80
(1973) 1099-1115.

Botts, Truman, *Probability theory and the* [4.3.2]
Lebesgue integral, Math. Magazine, 42 (1969)
105-111.

Chaitin, Gregory J., *Randomness and mathematical proof*, Scientific American, 232 (May 1975) 47-52, 122. [2.5.0]

David, F.N., Games, Gods and Gambling, Hafner, 1962. [1.2.0]

Doob, J.L., *Probability in function space*, Bull. Amer. Math. Soc., 53 (1947) 15-30. [4.3.2]

Hammersley, J.M., *Stochastic models for the distribution of particles in space*, Adv. Appl. Prob., 4 (1972) Suppl. pp. 44-68. [7.1.2]

Kingman, J.F.C., *Markov population processes*, J. Appl. Prob., 6 (1969) 1-18. [7.2.3]

Layzer, David, *The arrow of time*, Scientific American, 233 (December 1975) 56-69, 148. [7.1.1]

Mackey, George W., *Ergodic theory and its significance for statistical mechanics and probability theory*, Advances in Math., 12 (1974) 178-286. [4.7.3]

Maistrov, L.E., Probability Theory: A Historical Sketch, Academic Pr., 1974. [1.2.2]

Moran, P.A.P., *The probabilistic basis of stereology*, Adv. Appl. Prob., 4 (1972) Suppl. pp. 69-91. [7.6.2]

Ornstein, D.S., *Measure-preserving transformations and random processes*, Amer. Math. Monthly, 78 (1971) 833-840. [4.7.3]

Papanicolaou, G.C., *Stochastic equations and their applications*, Amer. Math. Monthly, 80 (1973) 526-545. [4.5.3]

Salmon, Wesley C., *Confirmation*, Scientific American, 228 (May 1973) 75-83, 120. [1.5.0]

Samuelson, Paul A., *Mathematics of speculative price*, SIAM Review, 15 (1973) 1-42. [7.4.2]

Sheynin, O.B., *On the prehistory of the theory of probability*, Arch. Hist. Exact Sci., 12 (1974) 97-141. [1.2.0]

6.2 STATISTICS

6.2.0 GENERAL

Adams, Ernest W., *On the nature and purpose of measurement*, <u>Synthese</u>, 16 (1966) 125-169; also in B. Lieberman, <u>Contemporary Problems in Statistics</u>, Oxford U. Pr., 1971, pp. 74-92.

Hotelling, Harold, *The statistical method and the philosophy of science*, <u>Amer. Statistician</u>, 12:5 (1958) 9-14.
> Distinguishes different notions of probability; discusses hypothesis testing, design of experiments, and sequential designs.

Kiefer, Jack, *Statistical inference*, in <u>The Mathematical Sciences--A Collection of Essays for COSRIMS</u>, M.I.T., 1969, pp. 60-71; also in <u>Math. Spectrum</u>, 3 (1970-1971) 1-11.
> Problem: A coin is known to be biased in the ratio two-to-one; it is tossed three times; devise a procedure for estimating whether it is biased in favor of heads or of tails. With this extraordinarily simple problem and certain of its modifications the author succeeds in illustrating a whole series of notions from the theory of statistical inference. In the process he manages to mention Abraham Wald, Karl Pearson, R.A. Fisher, Jerzy Neyman, Egon Pearson, and Thomas Bayes.

Kruskal, William, *Statistics, Molière, and Henry Adams*, <u>Amer. Scientist</u>, 55 (1967) 416-428.
> Attempts to give a feeling for the kinds of thinking and work that statisticians do in an engaging and delightful essay on methods for reaching conclusions from fallible data.

Moroney, M.J., <u>Facts from Figures</u>, Penguin, 1951.

Reichmann, W.J., <u>Use and Abuse of Statistics</u>, Penguin, 1964.

Tanur, Judith M., et al., <u>Statistics: A Guide to the Unknown</u>, Holden-Day, 1972.
> This unusual book contains 44 popular-level essays on a wide variety of subjects in which statistics is used, starting with a description of the 1954 Salk polio vaccine field trials. A number of leading figures in statistics and other fields have contributed articles. Mathematics and even arithmetic have been ruthlessly suppressed, so the articles can be read with enjoyment and comprehension at a very elementary level.

Weaver, Warren, *Statistics*, <u>Scientific American</u>, 186 (January 1952) 60-63, 84; also in M. Kline, <u>Mathematics in the Modern World</u>, Freeman, 1968, pp. 175-178, 399.

6.2.1 ELEMENTARY

Bancroft, T.A., *Statistics*, <u>Encyclopedia Americana</u>, 1976, V. 25, pp. 629-635.

Keeping, E.S., *Statistical decisions*, <u>Amer. Math. Monthly</u>, 63 (1956) 147-159.

Mood, A.M., *Statistics*, <u>McGraw-Hill Encyclopedia of Science and Technology</u>, 1960, V. 13, pp. 66-75.

Mosteller, Frederick and Hoaglin, David C., *Statistics*, <u>Encyclopaedia Britannica</u>, 15th ed., 1974, Macropaedia V. 17, pp. 615-624.

   Basic principles of statistical inference, estimation, and hypothesis testing are treated.

Savage, Leonard J., *The foundations of statistics reconsidered*, in H.E. Kyburg, Jr. and H.E. Smokler, <u>Studies in Subjective Probability</u>, Wiley, 1964, pp. 173-188.

Tukey, John W., *Mathematics and the picturing of data*, in <u>Proc. Inter. Cong. Math.</u> (1974), V. 2, Canad. Cong. Math., 1975, pp. 523-531.

6.2.2 ADVANCED

Kotz, Samuel and Johnson, Norman L., *Statistical distributions: a survey of the literature, trends and prospects*, <u>Amer. Statistician</u>, 27 (1973) 15-17.

   General remarks and a tabulation showing trends over the past 30 years in the use of some 40-odd specific statistical distributions, with references to the literature, including the authors' monumental four volume survey, <u>Distributions in Statistics</u> (Wiley, 1969, 1970, 1972).

Pratt, John W., Raiffa, Howard and Schlaifer, Robert, *The foundations of decision under uncertainty: an elementary exposition*, <u>J. Amer. Stat. Assoc.</u>, 59 (1964) 353-375.

Rosenkrantz, R.D., *The significance test controversy*, Synthese, 26 (1973) 304-321.

Careful discussion of the interpretation of exact significance levels and of the philosophical underpinnings of Neyman-Pearson theory.

Savage, Leonard J., *The theory of statistical decisions*, J. Amer. Stat. Assoc., 46 (1951) 55-67.

Smith, Cedric A.B., *Consistency in statistical inference and decision*, J. Royal Stat. Soc., (Ser. B) 23 (1961) 1-37.

6.2.3 RESEARCH

Birnbaum, Allan, *On the foundations of statistical inference*, J. Amer. Stat. Assoc., 57 (1962) 269-326.

Includes lengthy discussion by leading statisticians.

RELATED REFERENCES

Clark, R.M., *Statistics and radiocarbon dating*, Math. Spectrum, 7 (1974-75) 83-89. [7.6.0]

Clements, Forrest E., *Use of cluster analysis with anthropological data*, Amer. Anthropologist, 56 (1954) 180-199. [7.3.1]

Darrow, Karl K., *Memorial to the classical statistics*, Bell System Tech. J., 22 (1943) 108-135. [7.1.2]

Darrow, Karl K., *The new statistical mechanics*, Bell. System Tech. J., 22 (1943) 362-392. [7.1.2]

Good, I.J., *Analogues of Poisson's summation formula*, Amer. Math. Monthly, 69 (1962) 259-266. [6.4.3]

Grenander, Ulf, *Computational probability and statistics*, SIAM Review, 15 (1973) 134-192. [6.1.1]

Li, Ching Chun, *Biometrics*, McGraw-Hill Encyclopedia of Science and Technology, 1960, V. 2, pp. 223-232. [7.2.1]

Minlos, R.A., *Lectures on statistical physics*, Russian Math. Surveys, 23:1 (1968) 137-196. [7.1.2]

Robbins, Herbert, *Optimal stopping*, <u>Amer. Math. Monthly</u>, 77 (1970) 333-343. [6.1.2]

Schlaifer, Robert, *Expected value and utility*, in B. Lieberman, <u>Contemporary Problems in Statistics</u>, Oxford U. Pr., 1971, pp. 250-266. [7.4.0]

von Mises, Richard, <u>Probability, Statistics and Truth</u>, Macmillan, 1957. [6.1.0]

Wiener, Norbert, *The theory of prediction*, in E.F. Beckenbach, <u>Modern Mathematics for the Engineer</u>, McGraw-Hill, 1956, pp. 165-190. [4.7.3]

6.3 OPTIMIZATION

6.3.0 GENERAL

Bellman, Richard, *Dynamic programming treatment of the travelling salesman problem*, J. Assoc. Comp. Mach., 9 (1962) 61-63.

  Very simple algorithm, with discussion of computational feasibility in 1962 (17 cities possible).

Gale, David and Shapley, L.S., *College admissions and the stability of marriage*, Amer. Math. Monthly, 69 (1962) 9-15.

  Stable marriage by "deferred-acceptance." Tongue-in-cheek, yet mathematically significant. An excellent example of applied mathematics applied to the task of optimal assignment.

Kazarinoff, Nicholas D., Geometric Inequalities, New Math. Libr., No. 4, Random House, 1961; M.A.A., 1975.

  A very elementary yet sophisticated discussion of numerous inequalities dealing with geometric figures (e.g., isoperimetric theorems). The book closes with an elementary geometric proof of a theorem originally conjectured by Erdös and first proved by Mordel. Includes illustrative problems.

Rapoport, Anatol, *Critiques of game theory*, Behavioral Science, 4 (1959) 49-66.

  Analysis of criticism of game theory.

Rapoport, Anatol, *Escape from paradox*, Scientific American, 217 (July 1967) 50-56, 134.

  General historical discussion of paradoxes followed by game theory applied to the Prisoners' Dilemma.

Rapoport, Anatol, *The use and misuse of game theory*, Scientific American, 207 (December 1962) 108-118, 192; also in M. Kline, Mathematics in the Modern World, Freeman, 1968, pp. 304-312, 402-403; and in D. Messick, Mathematical Thinking in Behavioral Sciences, Freeman, 1968, pp. 95-103, 226.

  Fascinating examples ranging from poker to war.

Saaty, Thomas L., *Operations research: some contributions to mathematics*, Science, 178 (1972) 1061-1070.

  Survey of the nature and scope of operations research and how it draws on and contributes to various areas of mathematics.

6.3.1 ELEMENTARY

Beckenbach, Edwin F. and Bellman, Richard, An Introduction to Inequalities, New Math. Libr., No. 3, Random House, 1961; M.A.A., 1975.

> This book presents, at a very elementary level, what might be called the basic inequalities of arithmetic--if one includes in that category such results as the inequalities of Cauchy, Hölder, and Minkowski.

Bohnenblust, H. Frederic, The theory of games, in E.F. Beckenbach, Modern Mathematics for the Engineer, McGraw-Hill, 1956, pp. 191-210.

Chakerian, G.D. and Lange, L.H., Geometric extremum problems, Math. Magazine, 44 (1971) 57-69.

> Winner of the Lester R. Ford award for expository writing.

Dantzig, George B., Maximization of a linear function of variables subject to linear inequalities, in T.C. Koopmans, Activity Analysis of Production and Allocation, Wiley, 1951, pp. 339-347.

> The original paper on the simplex method for linear programming.

Dantzig, George B., On the shortest route through a network, Management Science, 6 (January 1960); also in D.R. Fulkerson, Studies in Graph Theory, Part I, M.A.A., 1975, pp. 89-93.

Fulkerson, D.R., Flow networks and combinatorial operations research, Amer. Math. Monthly, 73 (1966) 115-138; also in D.R. Fulkerson, Studies in Graph Theory, Part I, M.A.A., 1975, pp. 139-171.

> A flow network is a graph in which each edge is assigned an orientation and a non-negative number--its capacity--which may be thought of as the maximum rate at which it can transport a commodity. The article discusses the application of flow networks to a series of problems from operations research. Winner of the Lester R. Ford award for expository writing.

Gale, David, How to solve linear inequalities, Amer. Math. Monthly, 76 (1969) 589-599.

> The author describes methods for solving large systems of simultaneous linear equations and linear inequalities (the latter problem being closely related to linear programming). He is particularly interested in the efficacy of the methods in terms of the computation effort required, although he concentrates on theoretical upper bounds for this. "[T]here is a large and embarrassing gap between what has been observed and what has been proved. This gap... remains, in my opinion, the principal open question in the theory of linear computation."

Gale, David, *The jeep once more or jeeper by the dozen*, Amer. Math. Monthly, 77 (1970) 493-501; Correction, 78 (1971) 644-645.

The jeep problem is an optimization problem phrased in terms of a jeep which is to cross a desert by first making various partial trips in order to set up fuel depots--it can't carry enough fuel to make the trip directly. The author shows how to solve the problem and treats several generalizations.

Gomory, Ralph E., *Mathematical programming*, Amer. Math. Monthly, 72 (1965) Suppl. pp. 99-110.

A brief survey of linear, integer and dynamic programming, starting from the beginning. It would be a good source for a quick acquaintance with the field.

Hoffman, Alan J., *Linear programming*, McGraw-Hill Encyclopedia of Science and Technology, 1960, V. 7, pp. 522-523.

King, Gilbert W., *Applied mathematics in operations research*, in E.F. Beckenbach, Modern Mathematics for the Engineer, McGraw-Hill, 1956, pp. 211-242.

A very readable discussion of linear programming, Lagrange multipliers, game theory, the Poisson distribution, Monte Carlo methods, Bayes's theorem, prediction theory, and Markoff processes--all in the context of how they were being used in operations research in 1956. The author also engaged in some optimistic predictions: "...just as the scientific approach made chemistry out of alchemy, so it will make business a science, run by a new type of scientist."

Thompson, Gerald L., *Game theory*, McGraw-Hill Encyclopedia of Science and Technology, 1960, V. 6, pp. 24-30.

Tompkins, C.B., *Calculus of variations*, McGraw-Hill Encyclopedia of Science and Technology, 1960, V. 2, pp. 407-410.

Tucker, Albert W., *Combinatorial algebra of linear programs*, in J.G. Kemeny, R. Robinson and R.W. Ritchie, New Directions in Mathematics, Prentice-Hall, 1961, pp. 77-91; also in G.B. Dantzig and B.C. Eaves, Studies in Optimization, M.A.A., 1974, pp. 9-26.

The author takes as "a miniature example of linear programming" a simple explicit problem and solves it by various techniques. This gives the reader a quick entrée into such rudiments of the subject as duality and the simplex method.

Tucker, Albert W., et al., *Mathematical theory of optimization*, Encyclopaedia Britannica, 15th ed., 1974, Macropaedia V. 13, pp. 621-638.

Covers in four sections the theory of games, linear and non-linear programming, cybernetics, and control theory.

6.3.2 ADVANCED

Bellman, Richard, *Some applications of the theory of dynamic programming--a review*, Operations Research, 2 (1954) 275-288.

Cunningham, Frederic, Jr., *The Kakeya problem for simply connected and for star-shaped sets*, Amer. Math. Monthly, 78 (1971) 114-129.

  Winner of the Lester R. Ford award for expository writing.

Flanders, Harley, *A proof of Minkowski's inequality for convex curves*, Amer. Math. Monthly, 75 (1968) 581-593.

  Assuming of the reader only a modicum of knowledge of line integrals and of the Frenêt formulas, the author develops the treatment by calculus of convex curves in the plane. He then shows how to modify a method of Santaló so as to obtain Minkowski's inequality; the latter reduces in a special case to the classical isoperimetric inequality relating area to boundary length. An excellent starting place for one interested in learning this type of convexity theory. Winner of the Lester R. Ford award for expository writing.

Gomory, Ralph E. and Hu, T.C., *Multi-terminal flows in a network*, SIAM J. Appl. Math., 9 (1961) 551-570; also in D.R. Fulkerson, Studies in Graph Theory, Part I, M.A.A., 1975, pp. 172-199.

Good, R.A., *Systems of linear relations*, SIAM Review, 1 (1959) 1-31.

  Systems of linear inequalities, minimax theorem, linear programming, duality theory. Very neat development.

Karlin, Samuel, *The mathematical theory of inventory processes*, in E.F. Beckenbach, Modern Mathematics for the Engineer, 2nd Ser., McGraw-Hill, 1961, pp. 228-258.

  A few classical inventory models, such as the ones leading to "the famous square root formula", followed by several more sophisticated models.

Kuhn, Harold W., *"Steiner's" problem revisited*, in G.B. Dantzig and B.C. Eaves, Studies in Optimization, M.A.A., 1974, pp. 52-70.

  Steiner's problem is to minimize the sum of the distances from a point to the vertices of a triangle T. A dual geometric problem is to circumscribe T with an equilateral triangle of maximum altitude, and in the standard case the length of this altitude is the same as the original minimum sum. The author shows how these geometric facts generalize to a duality result in non-linear programming for which the dual has an interesting form, and gives an algorithm for solving the problem.

Lucas, William F., *An overview of the mathematical theory of games*, Management Science, 18 (1971-72) P3-P19.

Morse, Philip M., *Mathematical problems in operations research*, Bull. Amer. Math. Soc., 54 (1948) 602-621.

Nash, John, *Non-cooperative games*, Annals of Math., 54 (1951) 286-295.

Nash, John, *Two-person cooperative games*, Econometrica, 21 (1953) 128-140.

Powell, M.J.D., *A survey of numerical methods for unconstrained optimization*, SIAM Review, 12 (1970) 79-97.

> Unconstrained optimization is the process of finding a maximum value for a given function of n real variables, where each variable is unrestricted (i.e., the function is defined on all of n-space). Particular methods for accomplishing this numerically go back to Newton and Cauchy. This article surveys the refinements and new techniques which have been devised to improve efficiency when using a computer on such problems.

Vorob'ev, N.N., *The present state of the theory of games*, Russian Math. Surveys, 25:2 (1970) 77-136.

> An attempt to survey systematically the basic branches and directions of the subject. Not as extensive as the similar article by Lucas (cited above), nor as deep.

6.3.3 RESEARCH

Balinski, M.L., *Integer programming: methods, uses, computation*, Management Science, 12 (1965) 253-313.
> Presents the major methods plus computational experience.

Dubins, L.E. and Spanier, E.H., *How to cut a cake fairly*, Amer. Math. Monthly, 68 (1961) 1-17.
> Extensive bibliography.

Geoffrion, A.M. and Marsten, R.E., *Integer programming algorithms: a framework and state-of-the-art survey*, Management Science, 18 (1972) 465-491.

Payne, Lawrence F., *Isoperimetric inequalities and their applications*, SIAM Review, 9 (1967) 453-488.
> Winner of the LeRoy P. Steele prize for distinguished exposition of outstanding research.

Peterson, Elmer L., *Geometric programming*, <u>SIAM Review</u>, 18 (1976) 1-51.

Polak, E., *An historical survey of computational methods in optimal control*, <u>SIAM Review</u>, 15 (1973) 553-584.

Prager, W., *Mathematical programming and theory of structures*, <u>SIAM J. Appl. Math.</u>, 13 (1965) 312-332.
  Analogy between flows in networks and problems of plastic analysis, stress capacity of trusses, etc.

RELATED REFERENCES

Bellman, Richard, *Control theory*, <u>Scientific American</u>, 211 (September 1964) 186-200, 272; also in D. Messick, <u>Mathematical Thinking in Behavioral Sciences</u>, Freeman, 1968, pp. 74-82, 225. [7.6.0]

Berkovitz, Leonard D., *Optimal control theory*, <u>Amer. Math. Monthly</u>, 83 (1976) 225-239. [7.6.2]

Clark, Colin W., *The dynamics of commercially exploited natural animal populations*, <u>Math. Biosciences</u>, 13 (1972) 149-164. [7.2.1]

Courant, Richard, *Soap film experiments with minimal surfaces*, <u>Amer. Math. Monthly</u>, 47 (1940) 167-174. [5.3.1]

Eggleston, H.G., *The isoperimetric problem*, in N.J. Hardiman, <u>Exploring University Mathematics</u>, V. 1, Pergamon, 1967, pp. 95-120. [5.3.1]

Engel, Arthur, *The relevance of modern fields of applied mathematics for mathematical education*, <u>Educ. Studies Math.</u>, 2 (1969-70) 257-269. [1.6.1]

Hestenes, Magnus R., *An elementary introduction to the calculus of variations*, <u>Math. Magazine</u>, 23 (1950) 249-267. [4.7.1]

Hestenes, Magnus R., *Elements of the calculus of variations*, in E.F. Beckenbach, <u>Modern Mathematics for the Engineer</u>, McGraw-Hill, 1956, pp. 59-91. [4.7.2]

Krylov, V.I., *The calculus of variations*, [4.7.1]
in A.D. Aleksandrov, et al., Mathematics--
Its Content, Methods and Meaning, V. 2,
M.I.T., 1963, pp. 119-138.

Kuhn, Harold W. and Tucker, Albert W., [1.3.1]
*John von Neumann's work in the theory
of games and mathematical economics*,
Bull. Amer. Math. Soc., 64 (1958) Suppl.
pp. 100-122.

Rabinowitz, Philip, *Applications of linear* [6.4.2]
programming to numerical analysis, SIAM Review, 10 (1968) 121-159.

Saaty, Thomas L. and Alexander, Joyce M., [3.3.1]
*Optimization and the geometry of numbers:
packing and covering*, SIAM Review, 17 (1975)
475-519.

Shubik, Martin, *Games of status*, Behavioral [7.3.2]
Science, 16 (1971) 117-129.

Stein, Sherman K., *The mathematician as an* [1.5.0]
explorer, Scientific American, 204 (May 1961)
148-158, 206.

White, William W., *A status report on computing* [6.5.3]
algorithms for mathematical programming,
Computing Surveys, 5 (1973) 135-166.

6.4 NUMERICAL ANALYSIS

6.4.0 GENERAL

Bailey, D.E., *The effect of the solution of a problem of perturbations to the data: a study in interval arithmetic*, Math. Gazette, 57 (1973) 26-36.
  An adventure in approximation mathematics.

Brennan, Jean F., *Exploiting sparse matrices*, IBM Research Reports, 6:1 (1970) 1-8.

Davis, Philip J., *Numerical analysis*, in The Mathematical Sciences--A Collection of Essays for COSRIMS, M.I.T., 1969, pp. 128-137.
  A highly readable article which examines numerical analysis from many points of view (e.g., historical, educational, practical), and explores its relation to pure mathematics and to other computer sciences. "Numerical analysis therefore comprises the strategy of computation as well as the evaluation of what has been accomplished. The *beau ideal* of the subject (not often achieved, unfortunately) consists of 1) the formation of algorithms, 2) error analysis, including truncation and roundoff error, 3) the study of convergence including the rate of convergence, 4) comparative algorithms, which judges the relative utility of different algorithms in different situations."

Forsythe, George E., *Solving a quadratic equation on a computer*, in The Mathematical Sciences--A Collection of Essays for COSRIMS, M.I.T., 1969, pp. 138-152.
  The author describes floating point representation of numbers, points out its relationship to scientific notation, and discusses the problems that can arise in computer arithmetic from round-off error and overflow. He then uses the problem of finding the roots of a quadratic equation to illustrate pitfalls of numerical computation and methods which have been devised to avoid (some of) them, including so-called backward error analysis and an algorithm due to W. Kahan.

6.4.1 ELEMENTARY

Ashenhurst, R.L. and Metropolis, N., *Error estimation in computer calculation*, Amer. Math. Monthly, 72 (1965) Suppl. pp. 47-58.
  This article begins with some general remarks about the problem of error analysis, including definitions of three sources of error:

generated, inherent, and analytic. There follows discussion of such topics as error propagation and computer arithmetic. These ideas are illustrated by analyses of the determination of roots of polynomials and also of matrix pivoting.

Forsythe, George E., *Pitfalls in computation, or why a math book isn't enough*, Amer. Math. Monthly, 77 (1970) 931-956.

The author illustrates with a series of examples things that can go wrong when numerical mathematics is done poorly. Winner of the Lester R. Ford award for expository writing.

Householder, Alston S., *Generation of errors in digital computation*, Bull. Amer. Math. Soc., 60 (1954) 234-247.

Lowan, Arnold N., *Numerical analysis*, McGraw-Hill Encyclopedia of Science and Technology, 1960, V. 9, pp. 227-229.

Nikol'skiĭ, S.M., *Approximations of functions*, in A.D. Aleksandrov, et al., Mathematics--Its Content, Methods and Meaning, V. 2, M.I.T., 1963, pp. 265-302.

A review of basic notions: theorems dealing with interpolation polynomials, Tchebycheff polynomials, trigonometric polynomials, approximation of integrals, etc. It starts from the beginning and could serve as an excellent introduction to these topics.

Parlett, Beresford, *Matrix eigenvalue problems*, Amer. Math. Monthly, 72 (1965) Suppl. pp. 59-66.

In this article the author sets aside the problem of computing the eigenvalues of symmetric or Hermitian symmetric matrices on the grounds that highly satisfactory methods for these are known and describes instead methods for dealing with the asymmetric case.

6.4.2 ADVANCED

Brockett, R.W., *The synthesis of dynamical systems*, Quarterly of Applied Math., 30 (1972) 41-50.

The author discusses from a theoretical point of view the optimization of computational algorithms, using variations on Horner's rule for evaluating a polynomial as a model for much of the discussion. The title derives from the author's special way of thinking about these problems.

Glasser, M. Lawrence, *The summation of series*, SIAM J. Math. Anal., 2 (1971) 595-600.

A cure for slowly-converging series by rapidly converging integrals. Uses Laplace transforms.

Haber, Seymour, *Numerical evaluation of multiple integrals*, SIAM Review, 12 (1970) 481-526.

Numerical approximation of definite integrals is a subject with a rich history at least as old as calculus, but the first numerical method devised specifically for evaluating a multiple integral is attributed to James Clerk Maxwell in 1887, and real progress did not begin until the 1940's. This expository survey describes the various approximation theories which have been developed; the extent of the coverage is indicated by a bibliography of 123 items. Not intended as a practical guide to solving problems.

Householder, Alston S., *Numerical analysis*, in T.L. Saaty, Lectures on Modern Mathematics, V. 1, Wiley, 1963, pp. 59-97.

A leisurely introduction to various aspects of the numerical treatment of matrices.

Joyce, D.C., *Survey of extrapolation processes in numerical analysis*, SIAM Review, 13 (1971) 435-490.

"This survey traces the development of extrapolation processes in numerical analysis, dealing mainly with those based on polynomial or rational functions." The author lays the groundwork for this survey by describing the identical considerations which arise in calculating pi and in numerical integration, when the emphasis is on techniques for accelerating the convergence of approximations. With his notation and approach thus established, he discusses similar work in various areas of numerical analysis, giving attention to historical developments and providing voluminous references.

Nörlund, Niels E. and Mitchell, Andrew R., *Numerical analysis*, Encyclopaedia Britannica, 15th ed., 1974, Macropaedia V. 13, pp. 381-392.

The first section covers theoretical topics such as successive differences and interpolation formulas. The second section deals with various applications: approximation of functions, numerical differentiation and integration, numerical solutions to ordinary and partial differential equations.

Rabinowitz, Philip, *Applications of linear programming to numerical analysis*, SIAM Review, 10 (1968) 121-159.

After a very brief review of linear programming the author concentrates on a survey of the techniques for solving numerical analysis problems with the aid of linear programming.

Schoenberg, I.J., *The elementary cases of Landau's problem of inequalities between derivatives*, Amer. Math. Monthly, 80 (1973) 121-158.

Winner of the Lester R. Ford award for expository writing.

Strang, Gilbert, *The finite element method--linear and nonlinear applications*, in Proc. Inter. Cong. Math. (1974), V. 2, Canad. Cong. Math., 1975, pp. 429-435.

Tewarson, R.P., *Computations with sparse matrices*, SIAM Review, 12 (1970) 527-543.

    Matrices a large percentage of whose entries are zero are called sparse. Not only does sparseness simplify many calculations for large matrices, but, more important, various special methods are peculiarly appropriate to sparse matrices and greatly simplify calculations. This survey is primarily concerned with methods for inverting sparse matrices, although there are also brief comments on how to compute the eigenvalues of symmetric sparse matrices.

Varga, Richard S., *Iterative methods for solving matrix equations*, Amer. Math. Monthly, 72 (1965) Suppl. pp. 67-74.

    A simple one-dimensional problem is used to illustrate both the use of finite difference techniques to solve differential equations and the nature of iterative methods for solving the associated matrix equations.

Wilkinson, J.H., *Modern error analysis*, SIAM Review, 13 (1971) 548-568.

    While parts of this paper are technical, other parts can be read for general interest. These include the historical comments which center on a reappraisal of the classical paper of von Neumann and Goldstine giving error bounds for the inversion of positive definite matrices by the elimination method. "[T]heir results compare less unfavorably with recent error analyses than is generally supposed. The conscious adoption of backward error analysis and floating-point computation has greatly simplified error analysis..."

Young, David M., *A survey of modern numerical analysis*, SIAM Review, 15 (1973) 503-523.

6.4.3 RESEARCH

Good, I.J., *Analogues of Poisson's summation formula*, Amer. Math. Monthly, 69 (1962) 259-266.

    "The discrete Fourier transform is a stock-in-trade of the numerical analyst when he is engaged in practical Fourier analysis. It is not always recognized that it deserves a respected place in pure mathematics also. There are also applications to mathematical statistics, including the design of statistical experiments, and to the generation of random numbers. ... Our discussion will include a proof of a discrete analogue of Poisson's summation formula."

Lorentz, G.G., *Metric entropy, widths, and superpositions of functions*, Amer. Math. Monthly, 69 (1962) 469-485.

An excellent exposition of certain aspects of approximation theory. The author begins with theorems of D. Jackson and S. Bernstein concerning uniform approximation by trigonometric polynomials, and quickly moves on to relate these to the concepts of entropy and capacity, with emphasis on the work of Kolmogoroff. A point made to motivate the approach is that for practical reasons a programmer "will want to base his approximation methods on properties which are common to all functions of his class, without being interested to know that for some of them a better approximation exists."

RELATED REFERENCES

Birkhoff, Garrett, *Current trends in algebra*, Amer. Math. Monthly, 80 (1973) 760-782; Correction, 81 (1974) 746. [1.2.2]

Brennan, Jean F., *The fastest time of addition and multiplication*, IBM Research Reports, 4:1 (1968) 1-8. [6.5.1]

Polak, E., *An historical survey of computational methods in optimal control*, SIAM Review, 15 (1973) 553-584. [6.3.3]

Powell, M.J.D., *A survey of numerical methods for unconstrained optimization*, SIAM Review, 12 (1970) 79-97. [6.3.2]

Segel, Lee A., *The importance of asymptotic analysis in applied mathematics*, Amer. Math. Monthly, 73 (1966) 7-14. [4.7.3]

Ulam, Stanislaw M., *Monte Carlo calculations in problems of mathematical physics*, in E.F. Beckenbach, Modern Mathematics for the Engineer, 2nd Ser., McGraw-Hill, 1961, pp. 261-281. [6.1.2]

6.5 COMPUTER SCIENCE

6.5.0 GENERAL

Bauer, F.L., *Software and software engineering*, SIAM Review, 15 (1973) 469-480.

Birkhoff, Garrett, *Mathematics and computer science*, Amer. Scientist, 63 (1975) 83-91.

Forsythe, George E., *What to do till the computer scientist comes*, Amer. Math. Monthly, 75 (1968) 454-462.
> Long bibliography. Winner of the Lester R. Ford award for expository writing.

Goldstine, H., The Computer from Pascal to von Neumann, Princeton U. Pr., 1972.

Hamming, Richard W., *Impact of computers*, Amer. Math. Monthly, 72 (1965) Suppl. pp. 1-7.

Hamming, Richard W., *Intellectual implications of the computer revolution*, Amer. Math. Monthly, 70 (1963) 4-11; also in T.L. Saaty and F.J. Weyl, The Spirit and Uses of the Mathematical Sciences, McGraw-Hill, 1969, pp. 188-199; in Studies in Mathematics, V. 16, SMSG, 1967, pp. 45-52; and in Z.W. Pylyshyn, Perspectives on the Computer Revolution, Prentice-Hall, 1970, pp. 370-377.

Hamming, Richard W., *One man's view of computer science*, J. Assoc. Comp. Mach., 16 (1969) 3-12.
> Ethical dimension included.

Heath, F.G., *Origins of the binary code*, Scientific American, 227 (August 1972) 76-83, 124.
> Jacquard loom, Bacon cipher, Baudot telegraph.

Kac, Mark, *Will computers replace humans?*, in E.H. Kone and H.J. Jordan, The Greatest Adventure, Rockefeller U. Pr., 1974, pp. 193-206.

Kemeny, John G., *Mathematical models and the computer*, Pi Mu Epsilon Journal, 5 (1973) 373-386.
> Incisive speculation on the likely impact of computers on modeling in the coming decades. Suggests that models in the social sciences may run counter to an intuition honed on models in the physical sciences, and analyzes general advantages and disadvantages of computer modeling.

Knuth, Donald E., *Computer programming as an art*, <u>Comm. Assoc. Comp. Mach.</u>, 17 (1974) 667-673.

   Discussion of art vs. science, with conclusion that computer programming is both. Ends with a plea to computer scientists to "give us tools that are a pleasure to use."

Minsky, Marvin L., *Artificial intelligence*, <u>Scientific American</u>, 215 (September 1966) 246-260, 316; also in D. Messick, <u>Mathematical Thinking in Behavioral Sciences</u>, Freeman, 1968, pp. 141-148, 227.

   Interesting examples of machine intelligence--playing checkers, recognizing geometric analogies and communicating in English.

Minsky, Marvin L., *Form and content in computer science*, J. Assoc. Comp. Mach., 17 (1970) 197-215.

   Decries excessive formalism. Serious objections to the 'new math. Deals in detail with computer scientists' responsibility to education.

Perlis, Alan J., *Automatic programming*, <u>Quarterly of Applied Math.</u>, 30 (1972) 85-90.

   Brief remarks concerning the past and future of computer programming and artificial intelligence. "...automatic programming research will become absolutely intertwined with artificial intelligence work in this ensuing decade."

Randell, Brian, <u>The Origin of Digital Computers</u>, Springer-Verlag, 1973.

Robinson, Louis, et al., *Computers*, <u>Encyclopedia Americana</u>, 1976, V. 7, pp. 472-494.

Smith, Thomas M., *Some perspectives on the early history of computers*, in Z.W. Pylyshyn, <u>Perspectives on the Computer Revolution</u>, Prentice-Hall, 1970, pp. 7-15.

   An effective, nonlinear approach, emphasizing the convergence of major themes.

Zobrist, Albert L. and Carlson, Frederic R., Jr., *An advice-taking chess computer*, <u>Scientific American</u>, 228 (June 1973) 92-105, 124.

   A major departure from the vast "look-ahead" programs. (Chessboard diagrams incorporate novel use of second color to denote latest move made.)

6.5.1 ELEMENTARY

Adel'son-Vel'skii, G.M., et al., *Programming a computer to play chess*, Russian Math. Surveys, 25:2 (1970) 221-262.

   Includes algorithm for choosing the best move, description of heuristic methods, and how to structure a chess program. Nine sample games are given including four from a match with the U.S. computer at Stanford.

Brennan, Jean F., *The fastest time of addition and multiplication*, IBM Research Reports, 4:1 (1968) 1-8.

Knuth, Donald E., *Computer science and its relation to mathematics*, Amer. Math. Monthly, 81 (1974) 323-343.

   "My favorite way to describe computer science is to say that it is the study of *algorithms*." An extended treatment of "hashing"--storing and retrieving information--is used to illustrate the typical interplay between computer science and mathematics. Winner of the Lester R. Ford award for expository writing.

Knuth, Donald E., *Computer science and mathematics*, Amer. Scientist, 61 (1973) 707-713.

Kolata, Gina Bari, *Analysis of algorithms: coping with hard problems*, Science, 186 (1974) 520-521.

   Overview at a popular level of so-called NP-complete algorithms and their best probable approximations. (NP stands for non-deterministic polynomial time, and denotes an extensive class of combinatorial problems.)

Marimont, Rosalind B., *Applications of graphs and Boolean matrices to computer programming*, SIAM Review, 2 (1960) 259-268.

   Elementary. Discusses flow charts, transitive relations, powers of matrices.

Rankin, Bayard and Nelson, R.J., *Automata theory*, Encyclopaedia Britannica, 15th ed., 1974, Macropaedia V. 2, pp. 497-505.

Trakhtenbrot, B.A., *Algorithms*, in Z.W. Pylyshyn, Perspectives on the Computer Revolution, Prentice-Hall, 1970, pp. 69-86.

   Inviting introduction to algorithms, touching on games, Turing machines and Hilbert's tenth problem.

6.5.2 ADVANCED

Collins, G.E., *Computer algebra of polynomials and rational functions*, Amer. Math. Monthly, 80 (1973) 725-755.

> An article about computer science, written for mathematicians. The main topic is the algebraic manipulation of polynomials and rational functions in many variables (including polynomials over, e.g., the Gaussian integers), but there are necessary digressions into list processing and the time analysis of algorithms. Mathematical techniques range from the Euclidean algorithm to Hensel's p-adic lemma. Incidental intelligence: 200 digit integers are routinely dealt with in the programs.

Golomb, Solomon W. and Baumert, Leonard D., *Backtrack programming*, J. Assoc. Comp. Mach., 12 (1965) 516-524.

Greenspan, Donald, *Computer power and its impact on applied mathematics*, in A.A. Taub, Studies in Applied Mathematics, M.A.A., 1971, pp. 65-89.

> Several examples of computer analysis of problems in applied mathematics. One is the solution of a cavity flow problem by a nonlinear model, another is a mechanics problem in which the continuous model is discarded for a discrete model which is directly amenable to computer solution.

Robinson, J.A., *Theorem-proving on the computer*, J. Assoc. Comp. Mach., 10 (1963) 163-174.

> Survey article which contains several examples of proofs. Written at a high level--presupposes some sophistication in logic. Classifies levels of proofs.

Rogers, Hartley R., Jr., *The present state of Turing machine computability*, SIAM J. Appl. Math., 7 (1959) 114-130.

Schwartz, Jacob T., *Semantic and syntactic issues in programming*, Bull. Amer. Math. Soc., 80 (1974) 185-206.

> A lecture delivered as part of a special tutorial introduction to computer science.

Shepherdson, J.C. and Sturgis, H.E., *Computability of recursive functions*, J. Assoc. Comp. Mach., 10 (1963) 217-255.

> Recursive functions from a computer scientist's point of view, using machines a little more complicated than Turing machines.

6.5.3 RESEARCH

Lifshits, V.N. and Sadovskii, L.E., *Algebraic models of computing machines*, Russian Math. Surveys, 27:3 (1972) 87-135.

White, William W., *A status report on computing algorithms for mathematical programming*, Computing Surveys, 5 (1973) 135-166.
   Algorithms, data handling, and associated areas in linear programming and its extensions.

RELATED REFERENCES

Aberth, Oliver, *Analysis in the computable number field*, J. Assoc. Comp. Mach., 15 (1968) 275-299. [2.3.2]

Arbib, Michael A. and Manes, Ernest G., *Machines in a category: an expository introduction*, SIAM Review, 16 (1974) 163-192. [3.6.2]

Ashenhurst, R.L. and Metropolis, N., *Error estimation in computer calculation*, Amer. Math. Monthly, 72 (1965) Suppl. pp. 47-58. [6.4.1]

Birkhoff, Garrett, *Mathematics and psychology*, SIAM Review, 11 (1969) 429-469. [7.3.1]

Brockett, R.W., *The synthesis of dynamical systems*, Quarterly of Applied Math., 30 (1972) 41-50. [6.4.2]

Cannon, John J., *Computers in group theory: a survey*, Comm. Assoc. Comp. Mach., 12 (1969) 3-12. [3.5.3]

Chen, Wai-Kai, *Boolean matrices and switching nets*, Math. Magazine, 39 (1966) 1-8. [3.6.1]

Cohen, Hirsh, *Mathematical applications, computation, and complexity*, Quarterly of Applied Math., 30 (1972) 109-121. [1.4.1]

Dantzig, George B., *On the shortest route through a network*, Management Science, 6 (January 1960); also in D.R. Fulkerson, Studies in Graph Theory, Part I, M.A.A., 1975, pp. 89-93. [6.3.1]

Fromm, J.E. and Harlow, F.H., *Computer experiments in fluid dynamics*, Scientific American, 212 (March 1965) 104-110, 140. [7.1.0]

Geoffrion, A.M. and Marsten, R.E., *Integer programming algorithms: a framework and state-of-the-art survey*, Management Science, 18 (1972) 465-491. [6.3.3]

Glushkov, V.M., *The abstract theory of automata*, Russian Math. Surveys, 16:5 (1961) 1-54. [3.6.3]

Gordon, Richard, Herman, Gabor T. and Johnson, Steven A., *Image reconstruction from projections*, Scientific American, 233 (October 1975) 56-68, 139. [7.2.0]

Hall, Marshall, Jr. and Knuth, Donald E., *Combinatorial analysis and computers*, Amer. Math. Monthly, 72 (1965) Suppl. 21-28. [3.1.1]

Harlow, F.H., *Numerical fluid dynamics*, Amer. Math. Monthly, 72 (1965) Suppl. pp. 84-91. [7.1.1]

Hirolett, J., *Charles Babbage and his computer*, Math. Spectrum, 7 (1974-75) 73-80. [1.3.0]

Jones, James P., *Recursive undecidability--an exposition*, Amer. Math. Monthly, 81 (1974) 724-738. [2.5.2]

Knuth, Donald E., *Ancient Babylonian algorithms*, Comm. Assoc. Comp. Mach., 15 (1972) 671-677. [1.2.0]

Knuth, Donald E., *George Forsythe and the development of computer science*, Comm. Assoc. Comp. Mach., 15 (1972) 721-726. [1.3.0]

Lax, Peter D., *Numerical solution of partial differential equations*, Amer. Math. Monthly, 72 (1965) Suppl. pp. 74-84. [4.5.2]

Lehmer, D.H., *Computer technology applied to the theory of numbers*, in W.J. LeVeque, Studies in Number Theory, M.A.A., 1969, pp. 117-151. [3.3.2]

Lehmer, D.H., *Mechanized mathematics*, Bull. Amer. Math. Soc., 72 (1966) 739-750. [1.5.1]

Mellen, G.E., *Cryptology, computers, and*   [7.5.1]
common sense, Proc. Nat. Computer Conf., 42
(1973) 569-579.

Murray, Francis J. and Ford, Lester R.,   [1.5.0]
Mathematics as a calculatory science, Encyclopaedia Britannica, 15th ed., 1974, Macropaedia V. 11, pp. 671-696.

Rosen, Saul, *Electronic computers: a historical survey*, Computing Surveys, 1 (1969) 7-36.   [1.2.0]

7.1 Physical Science

7.1.0 GENERAL

Asimov, Isaac, *The ultimate speed limit*, Saturday Review, 55 (July 8, 1972) 53-56.

Brand, Louis, *The pi theorem of dimensional analysis*, Arch. Rat. Mech. Anal., 1 (1957) 34-45.

Drake, Stillman and MacLachlan, James, *Galileo's discovery of the parabolic trajectory*, Scientific American, 232 (March 1975) 102-110, 132.

Dyson, Freeman J., *Mathematics in the physical sciences*, Scientific American, 211 (September 1964) 128-146, 269-270; also in M. Kline, Mathematics in the Modern World, Freeman, 1968, pp. 249-257, 401; and in The Mathematical Sciences--A Collection of Essays for COSRIMS, M.I.T., 1969, pp. 97-115.

> The last half of the article discusses the use of group theory and the theory of group representations in research on fundamental (elementary) particles. This is introduced to illustrate the following point of view: "...in every century in which major advances were achieved the growth in physical understanding was guided by a combination of empirical observation with purely mathematical intuition. For a physicist mathematics is...the main source of concepts and principles by means of which new theories can be created."

Einstein, Albert, *On the generalized theory of gravitation*, Scientific American, 182 (April 1950) 13-17, 72; also in M. Kline, Mathematics in the Modern World, Freeman, 1968, pp. 258-262, 401.

> An elegant and motivated introduction to the "train of thought" that Einstein followed in arriving at the general theory of relativity.

Fromm, J.E. and Harlow, F.H., *Computer experiments in fluid dynamics*, Scientific American, 212 (March 1965) 104-110, 140.

Gardner, Martin, Relativity for the Million, Macmillan, 1962.

Gardner, Martin, The Ambidextrous Universe, Basic Books, 1964.

Lederberg, Joshua, *Topology of molecules*, in The Mathematical Sciences--A Collection of Essays for COSRIMS, M.I.T., 1969, pp. 37-51.

> In 1869 Jordan showed that any acyclic graph can be assigned a center. Indeed he showed that this can be done in either of two ways, yielding a so-called mass center or a radius center. The chemist selects one of these in order to assign a unique center to a complex molecule. This is a big first step in systematizing the description of molecules. Since the number of potential molecules is very large, a systematization--particularly one that can be put on a computer--is of considerable importance. This article describes some of the detail of the graph theory involved in this work. While not the best place to learn graph theory, it is an excellent place to get interested in graph theory.

Oxtoby, John C., *What are physical dimensions?*, Amer. Physics Teacher, 2 (Sept. 1934) 85-90.

Steen, Lynn Arthur, *Solving the great bubble mystery*, Science News, 108 (1975) 186-187.

7.1.1 ELEMENTARY

Aris, Rutherford, *Chemical kinetics and the ecology of mathematics*, Amer. Scientist, 58 (1970) 419-428.

> In this article linear algebra is used both to balance chemical reactions and to predict the products of reactions, while elementary differential equations are used to analyse chemical kinetics and the pseudo-steady-state hypothesis. "I believe mathematics must play a much more vital role for the engineer than just to provide him with a tool kit, and that at the same time its relationship with technological problems can provide a great deal of interest and stimulation to the mathematician."

Danby, John M.A., *Celestial mechanics*, Encyclopedia Americana, 1976, V. 6, pp. 125-127.

Friedrichs, K.O., From Pythagoras to Einstein, New Math. Libr., No. 16, Random House, 1965; M.A.A., 1975.

> Adopting as his theme the Pythagorean theorem, the author begins with the very elementary, including the algebra of vectors. In mid-book he shifts pace, slowly at first. Introducing momentum and energy, he discusses elastic and inelastic impact, without employing the notion of force. He then rediscusses these matters from the point of view of special relativity theory.

Hanson, Norwood Russell, *Number theory and physical theory: an analogy*, in R.S. Cohen and M.W. Wartofsky, <u>Boston Studies in the Philosophy of Science</u>, V. 2, Humanities Pr., 1965, pp. 93-119.

> Gödel applied to physics. Commentary by Armand Siegel follows on pp. 121-126.

Harlow, F.H., *Numerical fluid dynamics*, <u>Amer. Math. Monthly</u>, 72 (1965) Suppl. pp. 84-91.

> A general discussion of computer simulation of a fluid-dynamics situation, including problems involving shocks. Figures which look like photographs but are actually a variety of computer printout illustrate the success of the author's methods.

Kolata, Gina Bari, *Cascading bifurcations: the mathematics of chaos*, <u>Science</u>, 189 (1975) 984-985.

Layzer, David, *The arrow of time*, <u>Scientific American</u>, 233 (December 1975) 56-69, 148.

Lieber, Lillian R., <u>The Einstein Theory of Relativity</u>, Holt, Rinehart and Winston, 1936; 1945.

Rindler, W., *Survey of relativity theory*, <u>SIAM Review</u>, 3 (1961) 105-118.

> Elegant and elementary. Discusses Lorentz equations, Minkowski's principle.

Schild, Alfred, *The clock paradox in relativity theory*, <u>Amer. Math. Monthly</u>, 66 (1959) 1-18.

> Thought-provoking article on the geometry of relativity theory.

7.1.2 ADVANCED

Barr, Donald R., *When will the next record rainfall occur?*, <u>Math. Magazine</u>, 45 (1972) 15-19.

Binder, Raymond C., et al., *Mathematical aspects of physical theories*, <u>Encyclopaedia Britannica</u>, 15th ed., 1974, Macropaedia V. 14, pp. 392-424.

> The article deals with some fundamental mathematical aspects of theoretical physics. The theories discussed include statistical mechanics, electromagnetic theory, and quantum mechanics. Readers may be particularly interested in Einstein's section on Relativity theory (edited from the 13th edition of the <u>Britannica</u>).

Bolker, Ethan D., *The spinor spanner*, Amer. Math. Monthly, 80 (1973) 977-984.

To oversimplify, a rotation of 720° brings a "particle" back to where it started, but a rotation of 360° does not. This basic notion introduced into physics by Dirac is explained here from a topological point of view.

Brush, Stephen G., *Foundations of statistical mechanics 1845-1915*, Arch. Hist. Exact Sci., 4 (1967) 145-183.

Chandrasekhar, S., *Characteristic-value problems in hydrodynamic and hydromagnetic theory*, in E.F. Beckenbach, Modern Mathematics for the Engineer, 2nd Ser., McGraw-Hill, 1961, pp. 338-346.

The author discusses axisymmetric flow between two rotating concentric cylinders, and famous stability criteria due to Lord Rayleigh in the non-viscous case and G.I. Taylor in the viscous case. The emphasis is on the author's own work in analyzing how viscosity extends stability in the Taylor case beyond the "Rayleigh line." The details are meager, as the idea here is to give the background material and only to point out the key mathematical idea concerning differential equations.

Christie, Dan E., *Some thermodynamic properties of a mapping*, J. Franklin Inst., 267 (1959) 119-133.

Analysis of thermodynamic significance of properties of transformations of the plane. A beautiful synthesis of elementary physics and differential geometry.

Darrow, Karl K., *Memorial to the classical statistics*, Bell System Tech. J., 22 (1943) 108-135.

Darrow, Karl K., *The new statistical mechanics*, Bell System Tech. J., 22 (1943) 362-392.

Duff, G.F.D., *Mathematical problems of tidal energy*, in Proc. Inter. Cong. Math. (1974), V. 1, Canad. Cong. Math., 1975, pp. 87-94.

One percent of the earth's dissipation of tidal energy takes place in the Bay of Fundy and Gulf of Maine. If large scale construction took place in order to utilize this energy, it might have a major impact on the resonance phenomonon which creates the high tides involved. The author describes the mathematics which is used in attempts to analyze the behavior of the Fundy tides and to predict the impact of power plants that might some day be built.

Duffin, R.J., *Network models*, SIAM-AMS Proc., 3 (1971) 65-91; an expanded version appears as *Electrical network models* in D.R. Fulkerson, Studies in Graph Theory, Part I,

[7.1] Applications:

M.A.A., 1975, pp. 94-138.

> Some mathematicians have a geometric intuition, some see things best algebraically or analytically. A few think in terms of electrical networks. The author describes a variety of topics related to graph theory which can be illuminated from this point of view.

Dyson, Freeman, J., *Missed opportunities*, Bull. Amer. Math. Soc., 78 (1972) 635-652.

> Describes occasions in the past century when mathematicians and physicists failed to communicate with each other and thereby missed discoveries.

Frankel, Theodore, *Maxwell's equations*, Amer. Math. Monthly, 81 (1974) 343-349.

> The author proves a "folk-theorem" from physics using the formalism of exterior differential forms on Minkowski space. The theorem pertains to Maxwell's equations in a non-inductive medium, and states that the two more sophisticated ones (Faraday's law and the Ampere-Maxwell law) follow from the other two if one appeals to special relativity.

Hammersley, J.M., *Stochastic models for the distribution of particles in space*, Adv. Appl. Prob., 4 (1972) Suppl. pp. 44-68.

> Discusses laws of probability, definition of a Poisson process; introduces Markov fields and Gibbsian ensembles. Object is to describe physical phenomena arising from a random distribution of particles in space, especially on an atomic lattice in solid state physics. "...[T]he differences between one dimension and several dimensions are so great that the one-dimensional case is best thought of as a misleading anomaly foisted upon us by an understandable lack of mathematical expertise."

Kac, Mark, *Can one hear the shape of a drum?*, Amer. Math. Monthly, 73 (1966) Suppl. pp. 1-23.

> Deals with the equation of vibration of a membrane with clamped boundary--latitude being allowed in the selection of the boundary-- and the asymptotic behavior of the eigenvalues of this equation. The answer to the title question is known to be "yes" in the special case that the eigenvalues of the equation agree with the "overtones" of a circular drum. In the general case the answer is conjectured to be "no." Winner of the Lester R. Ford award and the Chauvenet Prize for expository writing.

Keller, Joseph B., *Inverse problems*, Amer. Math. Monthly, 83 (1976) 107-118.

> The author analyzes a number of inverse problems which have arisen in physics, e.g., an inverse scattering problem. He begins by

giving an interesting inverse problem posed and solved by Abel in 1826: "Suppose we slide a particle up a frictionless hill with initial energy E, and measure the time T(E) required for it to return. If we vary E and measure T(E), can we determine the shape of the hill from it?"

Klamkin, Murray S. and Newman, D.J., *The philosophy and applications of transform theory*, SIAM Review, 3 (1961) 10-36.

Starts with simple problems in arithmetic and geometry, then in probability, number theory, differential equations, ending with a boundary value problem. Transform, solve, invert. Beautiful lecture on heuristics in the Pólya tradition. Concludes with an exercise section.

Mackey, George W., *Quantum mechanics and Hilbert space*, Amer. Math. Monthly, 64 (1957) Suppl. pp. 45-57.

"[A]n attempt to give mathematicians who have not studied quantum mechanics some idea of what it is about and to give those who have studied it only from the point of view of the physicists some idea of how it may be formulated in a mathematically precise fashion... [While it is] based essentially on...von Neumann's classic book..., it differs...[in that] it is the writer's personal way of looking at the foundations of quantum mechanics."

MacLane, Saunders, *Hamiltonian mechanics and geometry*, Amer. Math. Monthly, 77 (1970) 570-586.

A concise, highly readable presentation of the elements of Hamiltonian mechanics in the language of differential manifolds along the lines expounded by Mackey, Smale, Abraham, Sternberg, and others. In this language, phase space, for example, becomes a cotangent bundle. In fact, many abstract mathematical notions such as vector bundles, K-theory, fibre bundles, etc., "were really discovered and used a long time ago in mechanics."

Minlos, R.A., *Lectures on statistical physics*, Russian Math. Surveys, 23:1 (1968) 137-196.

Six survey lectures.

Pekeris, C.L., *Adventures in applied mathematics*, Quarterly of Applied Math., 30 (1972) 67-83.

Brief descriptions of various problems in geophysics and atomic spectroscopy. The first example is "terrestial spectroscopy": under the impact of a severe earthquake the entire earth will pulsate at frequencies measured in cycles per hour; equations governing this pulsation are exhibited, and it is reported that their numerical solutions fit the experimental data very closely. Other geophysical problems discussed along similar lines are the dynamo theory of the origin of the earth's magnetic field, ocean tides, and seismic activity. In atomic spectroscopy, the author describes how he has calculated (with the aid of a massive recursion relation) the theoretical ionization energies of helium and ionized lithium.

Slater, J.C., *Physics and the wave equation*, Bull. Amer. Math. Soc., 52 (1946) 392-400.

Physicist's plea for mathematical work closer to physics. Josiah Willard Gibbs lecture.

Whitney, Hassler, *The mathematics of physical quantities*, Amer. Math. Monthly, 75 (1968) 115-138, 227-256.

Winner of the Lester R. Ford award for expository writing.

Wigner, Eugene P., *Symmetry principles in old and new physics*, Bull. Amer. Math. Soc., 74 (1968) 793-815.

7.1.3 RESEARCH

Carrier, George F., *Singular perturbation theory and geophysics*, SIAM Review, 12 (1970) 175-193; also in A.H. Taub, Studies in Applied Mathematics, M.A.A., 1971, pp. 1-26.

The paper begins with illuminating examples which convey some of the essence of singular perturbation theory. This theory is applied to the study of two geophysical problems: ocean circulation and the nature of hurricanes. Emphasis is on the interplay between mathematical and physical reasoning.

Cohen, Hirsh, *Nonlinear diffusion problems*, in A.H. Taub, Studies in Applied Mathematics, M.A.A., 1971, pp. 27-64.

An explanation of how the introduction of nonlinearity into the diffusion equation can drastically alter the qualitative behavior of the solution, e.g., by permitting wave fronts. The conduction of electrical impulses by nerve membranes is given as one of many examples. Cohen also describes the Stefan problem and its application to the propagation of phase transitions (as in the melting of ice) and to the switching of a superconducting material to the normal state.

Kato, Tosio, *Scattering theory*, in A.H. Taub, Studies in Applied Mathematics, M.A.A., 1971, pp. 90-115.

Lagerstrom, P.A. and Casten, R.G., *Basic concepts underlying singular perturbation techniques*, SIAM Review, 14 (1972) 63-120.

The authors' aim is "to show the fundamental heuristic ideas which underlie certain [singular perturbation] techniques" related to fluid dynamics. Emphasis is given to work of Kaplun and the understanding that this work contributes to the subject. While the paper becomes somewhat technical, the initial example of a damped vibrating string helps to convey some of the ideas.

Lin, C.C., *Dynamics of self-gravitating systems--structure of galaxies*, SIAM Review, 11 (1969) 127-151; also in A.H. Taub, Studies in Applied Mathematics, M.A.A., 1971, pp. 116-149.

The author is interested in how one explains mathematically the observed shapes of galaxies (two examples: spiral pattern, rotating bar). He also wishes to explain globular clusters, which are spherical sub-collections of stars within a galaxy (our galaxy has over a hundred globular clusters, each made up of hundreds of thousands of stars). The principal method described is an adaptation of the statistical treatment of gases, with the stars playing the role of individual molecules.

Mackie, A.G., *Some comments on existence and uniqueness theorems in applied mathematics with an application to thin airfoil theory*, SIAM Review, 10 (1968) 196-207.

The author gives several physical examples to illustrate his discussion of the judgments involved in idealizing an applied problem in order to set up appropriate differential equations describing it. He examines whether the resulting equations are "properly posed" and whether they satisfactorily describe the actual system. He discusses the Sommerfeld radiation condition in connection with a supersonic flow problem.

Taub, A.H., *Relativistic hydrodynamics*, in A.H. Taub, Studies in Applied Mathematics, M.A.A., 1971, pp. 150-180.

A survey for the reader who has some knowledge of classical hydrodynamics of how this subject can be modified to bring it into accord with the postulates of relativity theory.

Whitham, G.B., *Dispersive waves and variational principles*, in A.H. Taub, Studies in Applied Mathematics, M.A.A., 1971, pp. 181-212.

The author emphasizes the distinction between "hyperbolic" wave motion and "dispersive" wave motion, and concentrates on the latter--in which a local disturbance may disperse into a (usually changing) wave train. His illustrations include deep ocean swell from storms and deep ocean ship waves. Mathematically, emphasis is put on the role of the "dispersion relation" between frequency and wave number; in any particular problem this relation is derived from governing differential equations.

RELATED REFERENCES

Bergmann, Peter G., *Fifty years of relativity*, Science, 123 (1956) 487-494. [1.2.1]

Birkhoff, George D., *The mathematical nature of physical theories*, Amer. Scientist, 31 (1943) 281-310. [1.4.0]

Birkhoff, George D., *What is the ergodic theorem?*, Amer. Math. Monthly, 49 (1942) 222-226. [4.7.2]

Bochner, Salomon, *The role of mathematics in the rise of mechanics*, Amer. Scientist, 50 (1962) 294-311. [1.2.1]

Coxeter, H.S. MacDonald, *A geometrical background for de Sitter's world*, Amer. Math. Monthly, 50 (1943) 217-228. [5.1.1]

Coxeter, H.S. MacDonald, *The space-time continuum*, Historia Math., 2 (1975) 289-298. [1.2.1]

De Broglie, Louis, *The role of mathematics in the development of contemporary theoretical physics*, in F. LeLionnais, Great Currents of Mathematical Thought, V. 2, Dover, 1971, pp. 78-93. [1.4.0]

Drake, Stillman, *Galileo's discovery of the law of free fall*, Scientific American, 228 (May 1973) 84-92, 120. [1.2.1]

Drake, Stillman, *Mathematics and discovery in Galileo's physics*, Historia Math., 1 (1974) 129-150. [1.2.1]

Finkbeiner, Daniel T., *Vector and tensor analysis*, Encyclopaedia Britannica, 15th ed., 1974, Macropaedia V. 1, pp. 791-799. [3.6.2]

Gleason, Andrew M., *Evolution of an active mathematical theory*, Science, 145 (1964) 451-457; also appeared as *The evolution of differential topology*, in The Mathematical Sciences--A Collection of Essays for COSRIMS, M.I.T., 1969, pp. 176-189. [1.2.1]

Greenspan, Donald, *Computer power and its impact on applied mathematics*, in A.H. Taub, Studies in Applied Mathematics, M.A.A., 1971, pp. 65-89. [6.5.2]

Hersh, Reuben and Griego, Richard J., [6.1.1]
Brownian motion and potential theory,
Scientific American, 220 (March 1969) 66-74, 148.

Hoffman, Banesh, Albert Einstein: Creator [1.3.0]
and Rebel, Viking, 1972.

Kac, Mark, *On applying mathematics: reflections and examples*, Quarterly of Applied [1.5.3]
Math., 30 (1972) 17-29.

Kac, Mark, *Random walk and the theory of* [6.1.2]
Brownian motion, Amer. Math. Monthly, 54
(1947) 369-391.

Keller, Joseph B. and McLaughlin, D.W., [4.7.3]
The Feynman integral, Amer. Math. Monthly,
82 (1975) 451-465.

Krylov, V.I., *The calculus of variations*, [4.7.1]
in A.D. Aleksandrov, et al., Mathematics--
Its Content, Methods and Meaning, V. 2,
M.I.T., 1963, pp. 119-138.

Lanczos, Cornelius, Space Through the Ages, [5.1.1]
Academic Pr., 1970.

Lax, Peter D., *The formation and decay of* [4.5.2]
shock waves, Amer. Math. Monthly, 79 (1972)
227-241.

Lefschetz, Solomon, *Linear and nonlinear oscil-* [4.5.1]
lations, in E.F. Beckenbach, Modern Mathematics for the Engineer, McGraw-Hill, 1956, pp.
7-29.

Mackey, George W., *Ergodic theory and its* [4.7.3]
significance for statistical mechanics and
probability theory, Advances in Math., 12
(1974) 178-286.

Mackey, George W., *Group theory and its signi-* [3.5.0]
ficance for mathematics and physics, Proc.
Amer. Phil. Soc., 117 (1973) 374-380.

McShane, E.J., *Vector spaces and their ap-* [3.4.1]
plications, in The Mathematical Sciences--
A Collection of Essays for COSRIMS, M.I.T.,
1969, pp. 84-96.

Miles, John W., *Integral transforms*, in E.F. Beckenbach, <u>Modern Mathematics for the Engineer</u>, 2nd Ser., McGraw-Hill, 1961, pp. 68-99. [4.7.2]

Minty, G.J., *On the axiomatic foundations of the theories of directed linear graphs, electrical networks and network-programming*, <u>J. Math. and Mech.</u>, 15:3 (1966); also in D.R. Fulkerson, <u>Studies in Graph Theory</u>, Part II, M.A.A., 1975, pp. 246-300. [3.2.1]

Murnaghan, F.D., *An elementary presentation of the theory of quaternions*, <u>Scripta Math.</u>, 10 (1944) 37-49. [3.4.1]

Neményi, P.F., *The main concepts and ideas of fluid dynamics in their historical development*, <u>Arch. Hist. Exact Sci.</u>, 2 (1962) 52-86. [1.2.0]

Packel, Edward W., *Hilbert space operators and quantum mechanics*, <u>Amer. Math. Monthly</u>, 81 (1974) 863-873. [4.6.2]

Pólya, George, *Circle, sphere, symmetrization and some classical physical problems*, in E.F. Beckenbach, <u>Modern Mathematics for the Engineer</u>, 2nd Ser., McGraw-Hill, 1961, pp. 420-441. [4.7.2]

Prager, W., *Mathematical programming and theory of structures*, <u>SIAM J. Appl. Math.</u>, 13 (1965) 312-332. [6.3.3]

Randall, C.H. and Foulis, D.J., *An approach to empirical logic*, <u>Amer. Math. Monthly</u>, 77 (1970) 363-374. [2.6.1]

Salam, Abdus, *Theory of groups and the symmetry physicist*, <u>J. London Math. Soc.</u>, 41 (1966) 49-62. [5.8.3]

Shubnikov, A.V. and Koptsik, V.A., <u>Symmetry in Science and Art</u>, Plenum Pr., 1974. [3.5.1]

Sneddon, Ian N. and Smale, Stephen, *Differential equations*, <u>Encyclopaedia Britannica</u>, 15th ed., 1974, Macropaedia V. 5, pp. 736-767. [4.5.2]

Ulam, Stanislaw M., *Monte Carlo calculations in problems of mathematical physics*, in E.F. Beckenbach, <u>Modern Mathematics for the Engineer</u>, 2nd Ser., McGraw-Hill, 1961, pp. 261-281. [6.1.2]

Weyl, Hermann, *Relativity theory as a stimulus in mathematical research*, Proc. Amer. Phil. Soc., 93 (1949) 535-541. [1.2.2]

Weyl, Hermann, Symmetry, Princeton U. Pr., 1952; excerpted in J.R. Newman, The World of Mathematics, V. 1, Simon and Schuster, 1956, pp. 671-724. [1.4.1]

Wightman, A.S., *Analytic functions and elementary particles*, in The Mathematical Sciences-- A Collection of Essays for COSRIMS, M.I.T., 1969, pp. 116-127. [4.4.1]

Wigner, Eugene P., *The unreasonable effectiveness of mathematics in the natural sciences*, Comm. Pure Appl. Math., 13 (1960); also in T.L. Saaty and F.J. Weyl, The Spirit and Uses of the Mathematical Sciences, McGraw-Hill, 1969, pp. 123-140; in Studies in Mathematics, V. 16, SMSG, 1967, pp. 31-44; and in E.P. Wigner, Symmetries and Reflections: Scientific Essays of Eugene P. Wigner, Indiana U. Pr., 1967, pp. 222-237. [1.4.0]

Wilson, Curtis, *How did Kepler discover his first two laws?*, Scientific American, 226 (March 1972) 92-106, 126. [1.2.0]

Wyler, Oswald, *Exterior differential calculus and Maxwell's equations*, in K.O. May, Lectures on Calculus, Holden-Day, 1967, pp. 147-165. [4.2.2]

7.2 Biological Science

7.2.0 GENERAL

Bellman, Richard, *Mathematical models of the mind*, Math. Biosciences, 1 (1967) 287-304.
   Beautiful prose article; few formulas appear.

Bernhard, Robert, *Heresy in the halls of biology: mathematicians question Darwinism*, Scientific Research, (November 1967) 59-66.

Gordon, Richard, Herman, Gabor T. and Johnson, Steven A., *Image reconstruction from projections*, Scientific American, 233 (October 1975) 56-68, 139.

Moore, Edward F., *Mathematics in the biological sciences*, Scientific American, 211 (September 1964) 148-164, 270; also in M. Kline, Mathematics in the Modern World, Freeman, 1968, pp. 275-283, 401-402.
   Interesting introduction to population mathematics, genetics and the relation between abstract models and living organisms.

Rosen, Robert, *On mathematics and biology*, in T.L. Saaty and F.J. Weyl, The Spirit and Uses of the Mathematical Sciences, McGraw-Hill, 1969, pp. 203-218.

7.2.1 ELEMENTARY

Brearley, M.N., *The long jump miracle of Mexico City*, Math. Magazine, 45 (1972) 241-246.
   Did the altitude of Mexico City have any influence on the 1968 Olympic Games record-breaking long jump?

Clark, Colin W., *Profit maximization and the extinction of animal species*, J. Political Economy, 81 (1973) 950-961.
   Construction and analysis of a mathematical model for "commercial exploitation of a natural animal population." Concludes that "...extermination of the entire population may appear as the most attractive policy."

Clark, Colin W., *The dynamics of commercially exploited natural animal populations*, Math. Biosciences, 13 (1972) 149-164.

Cohen, Hirsh, *Mathematics and the biomedical sciences*, in The Mathematical Sciences--A Collection of Essays for COSRIMS, M.I.T., 1969, pp. 217-231.

> Historical illustrations together with informative, interesting insight into new areas of biomathematics. While advanced mathematics is occasionally required (e.g., in the use of differential equations to model the generation and propagation of nerve impulses) in many examples the role of mathematics is much more primitive.

Dantzig, George B., *New mathematical methods in the life sciences*, Amer. Math. Monthly, 71 (1964) 4-15.

> One model of gas exchange in lungs is given in detail; others are mentioned briefly.

Levins, Richard, *The strategy of model building in population biology*, Amer. Scientist, 54 (1966) 421-431.

> Examines the trade-offs among generality, realism, and precision, emphasizing the treatment of the same problem by means of alternative models with different simplifications but a common biological assumption. If several models lead to similar results, one may deduce a "robust theorem"--a conclusion relatively free of the details of the model. Several models are used to adduce the "robust theorem" that, "in an uncertain environment, species will evolve broad niches and tend toward polymorphism."

Li, Ching Chun, *Biometrics*, McGraw-Hill Encyclopedia of Science and Technology, 1960, V. 2, pp. 223-232.

Oxnard, Charles E., *Mathematics, shape and function: a study in primate anatomy*, Amer. Scientist, 57 (1969) 75-96.

> Numerical taxonomy and multivariate analysis give insight into the functional and evolutionary aspects of particular biological shapes.

Polachek, Harry, *The structure of the honeycomb*, Scripta Math., 7 (1940) 87-98.

Rashevsky, N., *Topology and life--in search of general mathematical principles in biology and sociology*, Bull. Math. Biophys., 16 (1954) 317-348.

> Well-written, interesting, introduction. Sets up graphs to represent different biological functions of an organism, such as digestion, movement, secretion of venom by a snake, etc. The problem is to find a proper transformation of a simple graph to graphs of higher organisms.

Zeeman, E. Christopher, *Catastrophe theory*, Scientific American, 234 (April 1976) 65-83, 138.

7.2.2 ADVANCED

Costello, W.G. and Taylor, H.M., *Deterministic population growth models*, Amer. Math. Monthly, 78 (1971) 841-855.

"A unified and systematic derivation of the classical demographic theory of long run deterministic population growth."

Hastings, S.P., *Some mathematical problems from neurobiology*, Amer. Math. Monthly, 82 (1975) 881-895.

"In this paper we discuss one of the few mathematical models which...give quantitatively accurate predictions of an important physiological process, the conduction of electrical impulses in a nerve axon. This model, which is due to the British physiologists A.E. Hodgkin and A.F. Huxley [in 1952],...is undoubtedly the most important mathematical model in neurobiology."

Herman, Gabor T. and Vitány, Paul M.B., *Growth functions associated with biological development*, Amer. Math. Monthly, 83 (1976) 1-15.

Hoffman, William C., *Visual illusions of angle as an application of Lie transformation groups*, SIAM Review, 13 (1971) 169-184.

The author first reviews some Lie group theory and then explains how he uses the Lie derivative to establish a principle for predicting optical illusions quantitatively. The principle is illustrated with applications to classical visual illusions: the Poggendorf illusion, "the apparent displacement of an oblique line which is broken by two parallel vertical lines", the Zollner illusion and the Hering illusion.

Kac, Mark, *Some mathematical models in science*, Science, 166 (1969) 695-699.

Karlin, Samuel, *Some mathematical models of population genetics*, Amer. Math. Monthly, 79 (1972) 699-739.

This is an excellent article for those who start with a good background in both genetics and mathematics. Winner of the Lester R. Ford award for expository writing.

Leslie, P.H., *On the use of matrices in certain population mathematics*, Biometrika, 33 (1945) 183-212.

Detailed projections of population distributions based on powers and spectral values of a transition matrix.

Lyubich, Yu. I., *Basic concepts and theorems of the evolutionary genetics of free populations*, Russian Math. Surveys, 26:5 (1971) 55-123.

"It is well known that the principles of biological inheritance, initiated by Mendel in 1865, allow of an exact mathematical formulation. For this reason classical genetics can be regarded as a mathematical discipline. This article is concerned with the direction in mathematical genetics that stems from the widely known papers of Hardy and Weinberg (1908)." An excellent example of the process of mathematical modelling, as results are always referred back to biological interpretations.

May, Robert M., *Biological populations with nonoverlapping generations: stable points, stable cycles, and chaos*, Science, 186 (1974) 645-647.

May, Robert M., *On relationships among various types of population models*, Amer. Naturalist, 107 (1973) 46-57.

Thom, René F., *Topological models in biology*, Topology, 8 (1969) 313-335.

One of the first papers on applications of catastrophe theory.

7.2.3 RESEARCH

Kingman, J.F.C., *Markov population processes*, J. Appl. Prob., 6 (1969) 1-18.

RELATED REFERENCES

Clark, Colin, *Some socially relevant applications of elementary calculus*, Two-Year College Mathematics Journal, 4 (1973) 1-15. [4.1.1]

Clark, Colin W., *The economics of overexploitation*, Science, 181 (1973) 630-634. [7.4.1]

Engel, Arthur, *The relevance of modern fields of applied mathematics for mathematical education*, Educ. Studies Math., 2 (1969-70) 257-269. [1.6.1]

Fejes Toth, L., *What the bees know and what they do not know*, Bull. Amer. Math Soc., 70 (1964) 468-481. [5.1.1]

Hoffer, William, *A magic ratio recurs throughout art and nature*, Smithsonian, 6 (December 1975) 110-124. [3.3.0]

Kendall, David G., *Branching processes since 1873*, J. London Math. Soc., 41 (1966) 385-406. [6.1.2]

Kendall, David G., *The genealogy of genealogy: branching processes before (and after) 1873*, <u>Bull. London Math. Soc.</u>, 7 (1975) 225-253. [6.1.2]

Panati, Charles, *Catastrophe theory*, <u>Newsweek</u> (January 19, 1976) 54-55. [7.3.0]

Stevens, Peter S., <u>Patterns in Nature</u>, Atlantic-Little, Brown, 1974. [5.1.0]

Stewart, Ian, *The seven elementary catastrophes*, <u>New Scientist</u> (29 November 1975) 447-454. [5.6.1]

Weyl, Hermann, <u>Symmetry</u>, Princeton U. Pr., 1952; excerpted in J.R. Newman, <u>The World of Mathematics</u>, V. 1, Simon and Schuster, 1956, pp. 671-724. [1.4.1]

Zeeman, E. Christopher, *Levels of structure in catastrophe theory illustrated by applications in the social and biological sciences*, in <u>Proc. Inter. Cong. Math.</u> (1974), V. 2, Canad. Cong. Math., 1975, pp. 533-546. [7.3.2]

7.3 BEHAVIORAL SCIENCE

7.3.0 GENERAL

Barbut, Marc, *Does the majority ever rule?*, Portfolio and Art News Annual, 4 (1961) 79-83, 161-168.
   An elementary exposition of Kenneth Arrow's theorem that the only decision procedure that satisfies certain elementary principles of social welfare is a dictatorship.

Kemeny, John G., *The social sciences call on mathematics*, in The Mathematical Sciences--A Collection of Essays for COSRIMS, M.I.T., 1969, pp. 21-36.
   A series of brief examples which illustrate areas of finite mathematics or its applications. Both the example for graph theory and the example for linear programming succeed not only in conveying basic notions of the subjects but in highlighting the use of a key theorem. Other examples describe computer simulation, a stripped-down ecological population-balance problem, and the use of Markov chains.

Panati, Charles, *Catastrophe theory*, Newsweek (January 19, 1976) 54-55.

7.3.1 ELEMENTARY

Balinski, M.L. and Young, H.P., *The quota method of apportionment*, Amer. Math. Monthly, 82 (1975) 701-730.
   A detailed historical account of the rules governing apportionment of the U.S. Congress revealing that known methods violate certain principles of fairness. Concludes with a (complex) new method that is then proved to be the unique method satisfying the essential properties.

Birkhoff, Garrett, *Mathematics and psychology*, SIAM Review, 11 (1969) 429-469.
   Discrete and continuous mathematical models of psychology, analysis of man-computer symbiosis and of psychological issues in artificial intelligence.

Clements, Forrest E., *Use of cluster analysis with anthropological data*, Amer. Anthropologist, 56 (1954) 180-199.
   A detailed description of how to do cluster analysis using the B-coefficient: the ratio between the average intercorrelations of a subset to the average correlation with the complementary set.

Although more sophisticated methods have been developed, this is an excellent illustration of what cluster analysis is all about and how it works.

Gulliksen, Harold, *Mathematical solutions for psychological problems*, Amer. Scientist, 47 (1959) 178-201.

Demonstrates the usefulness of matrix algebra and the generality of multidimensional scaling in psychological research.

Luce, R. Duncan, *The mathematics used in mathematical psychology*, Amer. Math. Monthly, 71 (1964) 364-378; also in Studies in Mathematics, V. 16, SMSG, 1967, pp. 181-195.

"The main issue in applying mathematics to psychological problems today, and most likely for some time to come, is the formulation of these problems in mathematical terms... . The main areas [to which mathematics has been applied] are those usually grouped together as 'experimental' psychology,...[a misnomer for] basic research into such fundamental psychological processes as learning, sensation, perception, and motivation." There follows a topical survey of such applications together with comments on their relation to various undergraduate courses. Winner of the Lester R. Ford award for expository writing.

Rapoport, Anatol, *Uses of mathematics outside the physical sciences*, SIAM Review, 15 (1973) 481-502.

Solomon, Herbert, *A survey of mathematical models in factor analysis*, in H. Solomon, Mathematical Thinking in the Measurement of Behavior, Free Pr., 1960, pp. 273-314.

7.3.2 ADVANCED

Boyd, John Paul, *The algebra of group kinship*, J. Math. Psych., 6 (1969) 139-167.

A study of kinship of certain kinds of societies utilizing group theory and relations. A strongly mathematical extension of some of the kinship models presented in H. White, An Anatomy of Kinship, Prentice-Hall, 1963, and new directions therefrom.

Krantz, David, *A survey of measurement theory*, in G.B. Dantzig and A.F. Veinott, Jr., Mathematics of the Decision Sciences, Part 2, Amer. Math. Soc., 1968, pp. 314-350.

Rapoport, Anatol, *Directions in mathematical psychology*, Amer. Math. Monthly, 83 (1976) 85-106, 153-172.

Rapoport, Anatol and Rebhun, L.I., *On the mathematical theory of rumor spread*, Bull. Math. Biophys., 14 (1952) 375-383.

Shubik, Martin, *Games of status*, Behavioral Science, 16 (1971) 117-129.

  Mathematical models are used to examine the results of nonconstant sum games in which a single value, status, is maximized.

Zeeman, E. Christopher, *Levels of structure in catastrophe theory illustrated by applications in the social and biological sciences*, in Proc. Inter. Cong. Math. (1974), V. 2, Canad. Cong. Math., 1975, pp. 533-546.

RELATED REFERENCES

Cohn, P.M., *Algebra and language theory*, Bull. London Math. Soc., 7 (1975) 1-29. [3.6.3]

Kemeny, John G., *Mathematical models and the computer*, Pi Mu Epsilon Journal, 5 (1973) 373-386. [6.5.0]

Saaty, Thomas L., *Operations research: some contributions to mathematics*, Science, 178 (1972) 1061-1070. [6.3.0]

Stewart, Ian, *The seven elementary catastrophes*, New Scientist (29 November 1975) 447-454. [5.6.1]

Zeeman, E. Christopher, *Catastrophe theory*, Scientific American, 234 (April 1976) 65-83, 138. [7.2.1]

Zeeman, E. Christopher, *The geometry of catastrophe*, Times Lit. Suppl., (10 December 1971) 1556-1557. [5.6.1]

7.4 ECONOMIC SCIENCE

7.4.0 GENERAL

Baumol, William J., *Mathematics in economic analysis*, in T.L. Saaty and F.J. Weyl, The Spirit and Uses of the Mathematical Sciences, McGraw-Hill, 1969, pp. 246-262.

Isard, Walter and Kaniss, Phyllis, *The 1973 Nobel prize for economic science*, Science, 182 (1973) 568-569, 571.
  Exposition of Wassily Leontief's development of the input-output matrix model for economic analysis.

Leontief, Wassily, *Mathematics in economics*, Bull. Amer. Math. Soc., 60 (1954) 215-233.
  Very general survey of modern economic theory, written by one of its principal architects.

Montgomery, David and Quirk, James, *Mathematics in economic theory*, SIAM News, 7:6 (1974) 2-3.
  Very general survey of Paul Samuelson and Kenneth Arrow's role in mathematicizing economics.

Schlaifer, Robert, *Expected value and utility*, in B. Lieberman, Contemporary Problems in Statistics, Oxford U. Pr., 1971, pp. 250-266.

7.4.1 ELEMENTARY

Clark, Colin W., *The economics of overexploitation*, Science, 181 (1973) 630-634.

Gale, David, *Mathematics and economic models*, Amer. Scientist, 44 (1956) 33-44.
  Simple linear and non-linear models of exchange, "jazzed-up" by introduction of variable prices and then of variable volumes. Good example of progressive stages of modelling.

Klein, Lawrence R., *The role of mathematics in economics*, in The Mathematical Sciences--A Collection of Essays for COSRIMS, M.I.T., 1969, pp. 161-175.

Samuelson, Paul A., *Maximum principles in analytical economics*, Science, 173 (1971) 991-997.
  In this Nobel Lecture the author draws various mathematical analogies between physical principles and economic principles.

7.4.2 ADVANCED

Debreu, Gerard, *Mathematical theory of economic equilibrium*, in <u>Proc. Inter. Cong. Math.</u> (1974), V. 1, Canad. Cong. Math., 1975, pp. 65-77.

> The purpose of this article is to introduce the mathematician to mathematical economics (particularly Walrasian economics). Thus the economics is held to a minimum, and the emphasis is on a clear exposition of the role of the mathematical tools. These include the Kakutani fixed point theorem, algorithms for calculating fixed points, the role of differential topology in establishing continuity for a set of equilibria, and the role of measure theory.

Gale, David, *A mathematical theory of optimal economic development*, <u>Bull. Amer. Math. Soc.</u>, 74 (1968) 207-223.

> A lucid introduction to growth economics written for mathematicians.

Gale, David, *On the theory of interest*, <u>Amer. Math. Monthly</u>, 80 (1973) 853-868.

> A simple analysis with surprisingly realistic results, that "involves nothing more exotic than separating a pair of convex sets by a hyperplane at the critical moment. This is the type of mathematics that has dominated applications in economics throughout the 'modern period' starting with Ville's proof of the von Neumann minimax theorem."

Koopmans, Tjalling C. and Bausch, Augustus F., *Selected topics in economics involving mathematical reasoning*, <u>SIAM Review</u>, 1 (1959) 79-148.

> Excellent survey with detailed references.

Samuelson, Paul A., *Mathematics of speculative price*, <u>SIAM Review</u>, 15 (1973) 1-42.

> This paper discusses stock warrant prices from the point of view of various aspects of probability theory. An appendix on continuous-time analysis by R.C. Merton (ibid., pp. 34-38) follows Samuelson's article.

7.4.3 RESEARCH

Mityagin, B.S., *Notes on mathematical economics*, <u>Russian Math. Surveys</u>, 27:3 (1972) 1-19.

> Two independent notes "...intended for mathematicians and economists who show a resigned optimism about the use of mathematical models in economics." Models of scientific and technical progress and Pontryagin's maximum principle; and structure of a set of equilibria in exchange models.

RELATED REFERENCES

Balinski, M.L., *Integer programming: methods, uses, computation*, Management Science, 12 (1965) 253-313. [6.3.3]

Dantzig, George B., *Maximization of a linear function of variables subject to linear inequalities*, in T.C. Koopmans, Activity Analysis of Production and Allocation, Wiley, 1951, pp. 339-347. [6.3.1]

Geoffrion, A.M. and Marsten, R.E., *Integer programming algorithms: a framework and state-of-the-art survey*, Management Science, 18 (1972) 465-491. [6.3.3]

Gomory, Ralph E. and Hu, T.C., *Multi-terminal flows in a network*, SIAM J. Appl. Math., 9 (1961) 551-570; also in D.R. Fulkerson, Studies in Graph Theory, Part I, M.A.A., 1975, pp. 172-199. [6.3.2]

Karlin, Samuel, *The mathematical theory of inventory processes*, in E.F. Beckenbach, Modern Mathematics for the Engineer, 2nd Ser., McGraw-Hill, 1961, pp. 228-258. [6.3.2]

Kuhn, Harold W. and Tucker, Albert W., *John von Neumann's work in the theory of games and mathematical economics*, Bull. Amer. Math. Soc., 64 (1958) Suppl. pp. 100-122. [1.3.1]

Lucas, William F., *An overview of the mathematical theory of games*, Management Science, 18 (1971-72) P3-P19. [6.3.2]

Nash, John, *Non-cooperative games*, Annals of Math., 54 (1951) 286-295. [6.3.2]

Nash, John, *Two-person cooperative games*, Econometrica, 21 (1953) 128-140. [6.3.2]

Rapoport, Anatol, *Critiques of game theory*, Behavioral Science, 4 (1959) 49-66. [6.3.0]

Rapoport, Anatol, *Escape from paradox*, Scientific American, 217 (July 1967) 50-56, 134. [6.3.0]

Rapoport, Anatol, *The use and misuse of game theory*, Scientific American, 207 (December 1962) 108-118, 192; also in M. Kline, Mathematics in the Modern World, Freeman, 1968, pp. 304-312, 402-403; and in D. Messick, Mathematical Thinking in Behavioral Sciences, Freeman, 1968, pp. 95-103, 226. [6.3.0]

Thompson, Gerald L., *Game theory*, McGraw-Hill Encyclopedia of Science and Technology, 1960, V. 6, pp. 24-30. [6.3.1]

7.5 INFORMATION SCIENCE

7.5.0 GENERAL

Gilbert, E.N., *An outline of information theory*, Amer. Statistician, 12:1 (1958) 13-19.

Hall, G.G., *An introduction to information theory*, Math. Teaching, 40 (1976) 4-7.

Peterson, W. Wesley, *Error-correcting codes*, Scientific American, 206 (February 1962) 96-108, 188; also in D. Messick, Mathematical Thinking in Behavioral Sciences, Freeman, 1968, pp. 52-58, 224.

7.5.1 ELEMENTARY

Gallager, Robert G., *Information theory*, Encyclopaedia Britannica, 15th ed., 1974, Macropaedia V. 9, pp. 574-580.
> Some of the topics included for discussion in this article are the measurement, encoding and transmission of information, Gaussian noise, and detection and estimation problems. Applications to cryptography and linguistics are discussed briefly.

Mellen, G.E., *Cryptology, computers, and common sense*, Proc. Nat. Computer Conf., 42 (1973) 569-579.
> Very good overall survey, particularly of recent developments in cryptology.

Reza, Fazlollah, *Information theory*, Encyclopedia Americana, 1976, V. 15, pp. 166-168.

Rogers, Hartley R., Jr., *Information theory*, Math. Magazine, 37 (1964) 63-78.
> Winner of the Lester R. Ford award for expository writing.

Sinkov, Abraham, Elementary Cryptanalysis--A Mathematical Approach, New Math. Libr., No. 22, Random House, 1968; M.A.A., 1975.
> Cryptanalysis is the art, or science, of breaking codes and ciphers which cryptography creates. (Cryptology embraces both disciplines.) The author's aim is to present elementary cryptanalysis from the mathematical point of view. The actual mathematics discussed is limited to quite elementary statistics and some simple facts about modular arithmetic and 2 by 2 matrices.

7.5.2 ADVANCED

Levinson, Norman, *Coding theory: a counterexample to G.H. Hardy's conception of applied mathematics*, Amer. Math. Monthly, 77 (1970) 249-258.
> The author reports on an area of applied mathematics where pure mathematics provides not only the theoretical background, but the actual constructive procedure. The area is coding theory, with emphasis on error correcting codes. The pure mathematics includes Galois fields and parts of number theory.

McMillan, Brockway, *An elementary approach to the theory of information*, SIAM Review, 3 (1961) 211-229.
> Based on finite probability spaces without Martingales or Radon-Nikodým derivatives. Develops Shannon's theory.

Varma, R.S. and Nath, Prem, *Information theory--a survey*, J. Math. Sci., 2 (1967) 75-109.

7.5.3 RESEARCH

Assmus, E.F., Jr. and Mattson, H.F., Jr., *Coding and combinatorics*, SIAM Review, 16 (1974) 349-388.

Elias, Peter, *The noisy channel coding theorem for erasure channels*, Amer. Math. Monthly, 81 (1974) 853-862.
> The Binary Erasure Channel (BEC) is a contrived version of a concept from information theory, the Binary Symmetric Channel (BSC). "The BEC is the pedagogical noisy channel par excellence. Results which are...difficult to prove for the BSC have analogs for the BEC which are obvious and which have transparent proofs. We give a selection of such results."

7.6 MISCELLANY

7.6.0 GENERAL

Bellman, Richard, *Control theory*, Scientific American, 211 (September 1964) 186-200, 272; also in D. Messick, Mathematical Thinking in Behavioral Sciences, Freeman, 1968, pp. 74-82, 225.

Brun, Viggo, *Euclidean algorithms and musical theory*, L'Enseignement Math., 10 (1964) 125-137.
> An application of the Euclidean algorithm as an algorithm of subtraction.

Budden, F.J., *Modern mathematics and music*, Math. Gazette, 51 (1967) 204-215.
> Analysis of musical scales.

Clark, R.M., *Statistics and radiocarbon dating*, Math. Spectrum, 7 (1974-75) 83-89.
> Shows how the statistician is able to help the archaeologist solve the latter's urgent problem of accurate carbon-14 dating.

Cohen, Joel E., *Information theory and music*, Behavioral Science, 7 (1962) 137-163.
> An extensive and ambitious essay.

Coxeter, H.S. MacDonald, *Music and mathematics*, Math. Teacher, 61 (1968) 312-320; also in Canad. Music J., 6 (1962) pp. 13-24.

Harris, Zellig, *Mathematical linguistics*, in The Mathematical Sciences--A Collection of Essays for COSRIMS, M.I.T., 1969, pp. 190-196.
> Extremely interesting essay on comparative linguistics--the study of common elements of human language and consequence insight into the nature of human thought. Although Harris uses some elementary mathematics in this essay, he opines that the contribution of mathematics to linguistics is "chiefly the attitude and the way of thinking rather than any particular result."

Kemeny, John G., *What every college president should know about mathematics*, Amer. Math. Monthly, 80 (1973) 889-901.
> "The most serious contribution that the mathematician-president can make is the fact that he knows something about model-building." The article is devoted to interesting models the author has had occasion to build.

Searle, John, *Chomsky's revolution in linguistics*, <u>N.Y. Review of Books</u>, 18 (June 29, 1972) 16-24.

Literate commentary on the evolution of linguistics under Noam Chomsky, including an annotated bibliography. Shows how Chomsky changed linguistics from a classification science to an axiomatic theory.

7.6.1 ELEMENTARY

Carter, F.L., *Perspective drawing by numbers*, <u>Math. Gazette</u>, 53 (1969) 133-139.

Use of matrices in the problem of constructing an accurate perspective drawing.

Crowe, Donald W., *The geometry of African art I. Bakuba art*, <u>J. of Geometry</u>, 1 (1971) 169-182; *II. A catalog of Benin patterns*, <u>Historia Math.</u>, 2 (1975) 253-271.

Classification of repeated patterns in African art on the basis of the 24 plane crystallographic groups; includes drawings.

Roberts, Fred S. and Brown, Thomas A., *Signed digraphs and the energy crisis*, <u>Amer. Math. Monthly</u>, 82 (1975) 577-594.

Describes a graph-theoretic method of modeling complex systems, such as the energy crisis, where there exists only a minimal amount of information about the system.

Segel, Lee A., *Simplification and scaling*, <u>SIAM Review</u>, 14 (1972) 547-571.

7.6.2 ADVANCED

Berkovitz, Leonard D., *Optimal control theory*, <u>Amer. Math. Monthly</u>, 83 (1976) 225-239.

Cowan, Thaddeus M., *The theory of braids and the analysis of impossible figures*, <u>J. Math. Psych.</u>, 11 (1974) 190-212.

A systematic way of generating impossible figures, using Artin's theory of braids.

Haggett, Peter, *Network models in geography*, in R.J. Chorley and P. Haggett, <u>Models in Geography</u>, Methuen, 1967, pp. 609-668.

Excellent survey article of uses of path geometry, tree geometry, circuit geometry, cell geometry, and network transformations in geography. Large bibliography.

Lambek, Joachim, *The mathematics of sentence structures*, Amer. Math. Monthly, 65 (1958) 154-170.

Develops an algorithm for distinguishing sentences from non-sentences. Both formal and natural languages. Includes Gentzen's decision procedure for propositional calculus.

Moran, P.A.P., *The probabilistic basis of stereology*, Adv. Appl. Prob., 4 (1972) Suppl. pp. 69-91.

"Stereology is the subject which is concerned with the mensuration of three-dimensional aggregates of materials by estimates based on two-dimensional (flat) sections, or by one-dimensional traverses (straight lines) through the aggregate." Various measures of particle size and the estimation of size distribution of spheres are described. Concludes with a discussion of beer foam as a model for division of space, and applications to metallurgy.

Oettinger, A.G., *Computational linguistics*, Amer. Math. Monthly, 72 (1965) Suppl. pp. 147-150.

Long bibliography.

Spanier, E.H., *Grammars and languages*, Amer. Math. Monthly, 76 (1969) 335-342.

RELATED REFERENCES

Nalimov, V.V., *Logical foundations of applied mathematics*, Synthese, 27 (1974) 211-250. [1.5.2]

Stone, Marshall H., *Mathematics and the future of science*, Bull. Amer. Math. Soc., 63 (1957) 61-76. [1.4.0]

INDEX BY AUTHOR

Each bibliographic entry listed in this Index is followed by a three-digit code and a page number that indicate the location of the primary entry in the bibliography. Articles that have more than one author are listed in their entirety under each of their authors.

Aaboe, Asger, <u>Episodes From the Early History of Mathematics</u>, New Math. Libr., No. 13, Random House, 1964; M.A.A., 1975. [1.2.1]; p. 6.

Abbott, Edwin A., <u>Flatland--A Romance of Many Dimensions</u>, Little, Brown, 1928; Dover, 1952. [1.7.0]; p. 36.

Aberth, Oliver, *Analysis in the computable number field*, <u>J. Assoc. Comp. Mach.</u>, 15 (1968) 275-299. [2.3.2]; p. 48.

Abian, A., *On inaccessible cardinal numbers*, <u>Arch. Math. Logik</u>, 12 (1969) 99-103. [2.1.2]; p. 39.

———, *The Stone space of a Boolean ring*, <u>L'Enseignement Math.</u>, 11 (1965) 194-198. [3.6.2]; p. 85.

Adams, Ernest W., *On the nature and purpose of measurement*, <u>Synthese</u>, 16 (1966) 125-169; also in B. Lieberman, <u>Contemporary Problems in Statistics</u>, Oxford U. Pr., 1971, pp. 74-92. [6.2.0]; p. 161.

Adel'son-Vel'skii, G.M., et al., *Programming a computer to play chess*, <u>Russian Math. Surveys</u>, 25:2 (1970) 221-262. [6.5.1]; p. 179.

Adler, Alfred, *Reflections--mathematics and creativity*, <u>New Yorker</u>, 47 (February 19, 1972) 39-45. [1.5.0]; p. 25.

Agazzi, Evandro, *The rise of the foundational research in mathematics*, <u>Synthese</u>, 27 (1974) 7-26. [2.2.0]; p. 43.

Ahlfors, Lars V., *Quasiconformal mappings and their applications*, in T.L. Saaty, <u>Lectures on Modern Mathematics</u>, V. 2, Wiley, 1964, pp. 151-164. [4.4.3]; p. 107.

Alder, Henry L., *Partition identities--from Euler to the present*, <u>Amer. Math. Monthly</u>, 76 (1969) 733-746. [3.1.2]; p. 61.

Aleksandrov, A.D., *Curves and surfaces*, in A.D. Aleksandrov, et al., <u>Mathematics--Its Content, Methods and Meaning</u>, V. 2, M.I.T., 1963, pp. 57-117. [5.3.1]; p. 133.

———, *Non-euclidean geometry*, in A.D. Aleksandrov, et al., <u>Mathematics--Its Content, Methods and Meaning</u>, V. 3, M.I.T., 1963, pp. 97-189. [5.1.1]; p. 126.

Aleksandrov, P.S., *On some basic directions in general topology*, <u>Russian Math. Surveys</u>, 19:6 (1964) 1-40. [5.5.3]; p. 142.

———, *Poincaré and topology*, <u>Russian Math. Surveys</u>, 27:1 (1972) 157-168. [5.5.2]; p. 141.

A

———, *Some results in the theory of topological spaces, obtained within the last 25 years*, <u>Russian Math. Surveys</u>, 15:2 (1960) 23-84. [5.5.2]; p. 141.

———, *Topology*, in A.D. Aleksandrov, et al., <u>Mathematics--Its Content, Methods and Meaning</u>, V. 3, M.I.T., 1963, pp. 193-225. [5.5.1]; p. 139.

Alexander, Joyce M. and Saaty, Thomas L., *Optimization and the geometry of numbers: packing and covering*, <u>SIAM Review</u>, 17 (1975) 475-519. [3.3.1]; p. 69.

Algebraic geometry, <u>Encyclopaedia Britannica</u>, 15th ed., 1974, Macropaedia V. 7, pp. 1070-1076. [5.4.3]; p. 137.

Algebraic topology, <u>Encyclopaedia Britannica</u>, 15th ed., 1974, Macropaedia V. 18, pp. 504-509. [5.7.2]; p. 149.

Allan, G.R., *Some aspects of the theory of commutative Banach algebras and holomorphic functions of several complex variables*, <u>Bull. London Math. Soc.</u>, 3 (1971) 1-17. [4.6.3]; p. 117.

Allendoerfer, Carl B., *Generalizations of theorems about triangles*, <u>Math. Magazine</u>, 38 (1965) 253-259. [5.1.1]; p. 126.

Almgren, F.J., Jr. and Montgomery, H., *The 1974 Fields Medals (II): An analyst and number theorist*, <u>Science</u>, 186 (1974) 130-131. [1.3.3]; p. 20.

Andrus, Jan F. and Butson, Alton T., *Ordered groups*, <u>Amer. Math. Monthly</u>, 70 (1963) 619-628. [3.5.2]; p. 80.

Arbib, Michael A. and Manes, Ernest G., *Machines in a category: an expository introduction*, <u>SIAM Review</u>, 16 (1974) 163-192. [3.6.2]; p. 85.

Archibald, R.C., *The first translation of Euclid's Elements into English and its source*, <u>Amer. Math. Monthly</u>, 57 (1950) 443-452. [5.1.0]; p. 125.

Archibald, Ralph G., *Goldbach's theorem*, <u>Scripta Math.</u>, 3 (1935) 44-50, 153-161. [3.3.1]; p. 68.

———, *Waring's problem: squares*, <u>Scripta Math.</u>, 7 (1940) 33-48. [3.3.2]; p. 69.

Aris, Rutherford, *Chemical kinetics and the ecology of mathematics*, <u>Amer. Scientist</u>, 58 (1970) 419-428. [7.1.1]; p. 185.

Arnold, B.H., *Boolean algebra*, <u>Encyclopedia Americana</u>, 1976, V. 4, pp. 254-255. [3.6.1]; p. 84.

———, *Set theory*, <u>Encyclopedia Americana</u>, 1976, V. 24, pp. 588-591. [2.1.1]; p. 38.

Artin, Emil, *The theory of braids*, <u>Amer. Scientist</u>, 38 (1950) 112-119; also in <u>Math. Teacher</u>, 52 (1959) 328-333. [3.5.0]; p. 79.

Ashenhurst, R.L. and Metropolis, N., *Error estimation in computer calculation*, <u>Amer. Math. Monthly</u>, 72 (1965) Suppl. pp. 47-58. [6.4.1]; p. 172.

Asimov, Isaac, *The ultimate speed limit*, Saturday Review, 55 (July 8, 1972) 53-56. [7.1.0]; p. 184.

Assmus, E.F., Jr. and Mattson, H.F., Jr., *Coding and combinatorics*, SIAM Review, 16 (1974) 349-388. [7.5.3]; p. 209.

Atiyah, M.F., *The role of algebraic topology in mathematics*, J. London Math. Soc., 41 (1966) 63-69. [5.7.2]; p. 149.

Ayer, A.J., *Chance*, Scientific American, 213 (October 1965) 44-54, 126-127; also in M. Kline, Mathematics in the Modern World, Freeman, 1968, pp. 151-160, 398; and in D. Messick, Mathematical Thinking in Behavioral Sciences, Freeman, 1968, pp. 4-13, 223. [6.1.1]; p. 155.

Ayoub, Raymond, *Euler and the zeta function*, Amer. Math. Monthly, 81 (1974) 1067-1086; Addendum, 82 (1975) 737. [1.2.2]; p. 9.

Bailey, D.E., *The effect of the solution of a problem of perturbations to the data: a study in interval arithmetic*, Math. Gazette, 57 (1973) 26-36. [6.4.0]; p. 172.

Bailey, Herbert S., Jr. and Tucker, Albert W., *Topology*, Scientific American, 182 (January 1950) 18-24, 64; also in M. Kline, Mathematics in the Modern World, Freeman, 1968, pp. 134-140, 398. [5.5.0]; p. 139.

Balinski, M.L., *Integer programming: methods, uses, computation*, Management Science, 12 (1965) 253-313. [6.3.3]; p. 169.

────── and Young, H.P., *The quota method of apportionment*, Amer. Math. Monthly, 82 (1975) 701-730. [7.3.1]; p. 201.

Ball, W.W. Rouse and Coxeter, H.S. MacDonald, Mathematical Recreations & Essays, Twelfth Edition, U. of Toronto Pr., 1974. [1.7.0]; p. 36.

Ballier, F. and Köthe, G., *The changing structure of modern mathematics*, in H. Behnke, et al., Fundamentals of Mathematics, V. 3, M.I.T., 1974, pp. 505-528. [1.5.2]; p. 30.

Banchoff, Thomas F., *Critical points and curvature for embedded polyhedral surfaces*, Amer. Math. Monthly, 77 (1970) 475-485. [5.3.2]; p. 133.

Bancroft, T.A., *Statistics*, Encyclopedia Americana, 1976, V. 25, pp. 629-635. [6.2.1]; p. 162.

Barbut, Marc, *Does the majority ever rule?*, Portfolio and Art News Annual, 4 (1961) 79-83, 161-168. [7.3.0]; p. 201.

Barker, Stephen F., *Realism as a philosophy of mathematics*, in J.J. Bulloff, T.C. Holyoke and S.W. Hahn, Foundations of Mathematics, Springer-Verlag, 1969, pp. 1-9. [2.2.0]; p. 43.

Barnard, Raymond W., et al., *Analytic and trigonometric geometry*, Encyclopedia Britannica, 15th ed., 1974, Macropaedia V. 7, pp. 1076-1093. [5.1.0]; p. 125.

Baron, Margaret E., *The Origins of the Infinitesimal Calculus*, Pergamon, 1969. [1.2.1]; p. 6.

Barr, Donald R., *When will the next record rainfall occur?*, Math. Magazine, 45 (1972) 15-19. [7.1.2]; p. 186.

Barr, Stephen, *Experiments in Topology*, Thomas Crowell, 1964. [5.5.0]; p. 139.

Barwise, Jon, *Back and forth through infinitary logic*, in M.D. Morley, Studies in Model Theory, M.A.A., 1973, pp. 5-34. [2.6.3]; p. 57.

Batchelder, P.M., *Waring's problem*, Amer. Math. Monthly, 43 (1936) 21-27. [3.3.1]; p. 68.

Bateman, Paul T. and Diamond, Harold G., *Asymptotic distribution of Beurling's generalized prime numbers*, in W.J. LeVeque, Studies in Number Theory, M.A.A., 1969, pp. 152-210. [3.3.3]; p. 72.

Bauer, F.L., *Software and software engineering*, SIAM Review, 15 (1973) 469-480. [6.5.0]; p. 177.

Baum, Robert J., Philosophy and Mathematics, Freeman Cooper, 1973. [2.2.0]; p. 43.

Baumert, Leonard D. and Golomb, Solomon W., *Backtrack programming*, J. Assoc. Comp. Mach., 12 (1965) 516-524. [6.5.2]; p. 180.

Baumol, William J., *Mathematics in economic analysis*, in T.L. Saaty and F.J. Weyl, The Spirit and Uses of the Mathematical Sciences, McGraw-Hill, 1969, pp. 246-262. [7.4.0]; p. 204.

Bausch, Augustus F. and Koopmans, Tjalling C., *Selected topics in economics involving mathematical reasoning*, SIAM Review, 1 (1959) 79-148. [7.4.2]; p. 205.

Baylis, E.R., *Knots--a practical application of group theory*, Math. Gazette, 57 (1973) 311-320. [3.5.2]; p. 80.

Beaumont, Ross A., *Determinants*, McGraw-Hill Encyclopedia of Science and Technology, 1960, V. 4, pp. 81-84. [3.4.1]; p. 75.

————, *Linear systems of equations*, McGraw-Hill Encyclopedia of Science and Technology, 1960, V. 7, pp. 523-525. [3.4.1]; p. 75.

————, *Matrix theory*, McGraw-Hill Encyclopedia of Science and Technology, 1960, V. 8, pp. 182-184. [3.4.1]; p. 75.

————, *Theory of equations*, McGraw-Hill Encyclopedia of Science and Technology, 1960, V. 5, pp. 46-47. [3.6.1]; p. 84.

Beckenbach, Edwin F., *Conformal mapping methods*, in E.F. Beckenbach, Modern Mathematics for the Engineer, McGraw-Hill, 1956, pp. 361-388. [4.4.2]; p. 105.

———— and Bellman, Richard, An Introduction to Inequalities, New Math. Libr., No. 3, Random House, 1961; M.A.A., 1975. [6.3.1]; p. 166.

Belinfante, Johan G.F. and Kolman, Bernard, *An introduction to Lie groups and Lie algebras, with applications, I-III*, SIAM Review, 8 (1966) 11-46; 10 (1968) 160-195; 11 (1969) 510-543; revised and reprinted as A Survey of Lie Groups and Lie Algebras, SIAM, 1972. [5.8.3]; p. 152.

Bell, Eric Temple, *Gauss and the early development of algebraic numbers*, National Math. Magazine, 18 (1944) 188-204, 219-233. [1.2.2]; p. 10.

———, Men of Mathematics, Simon and Schuster, 1937. [1.3.0]; p. 17.

———, *Newton after three centuries*, Amer. Math. Monthly, 49 (1942) 553-575. [1.2.1]; p. 6.

———, The Development of Mathematics, McGraw-Hill, 1945. [1.2.1]; p. 6.

Bellman, Richard, *Control theory*, Scientific American, 211 (September 1964) 186-200, 272; also in D. Messick, Mathematical Thinking in Behavioral Sciences, Freeman, 1968, pp. 74-82, 225. [7.6.0]; p. 210.

———, *Dynamic programming treatment of the travelling salesman problem*, J. Assoc. Comp. Mach., 9 (1962) 61-63. [6.3.0]; p. 165.

———, *Mathematical models of the mind*, Math. Biosciences, 1 (1967) 287-304. [7.2.0]; p. 196.

———, *Some applications of the theory of dynamic programming--a review*, Operations Research, 2 (1954) 275-288. [6.3.2]; p. 168.

——— and Beckenbach, Edwin F., An Introduction to Inequalities, New Math. Libr., No. 3, Random House, 1961; M.A.A., 1975. [6.3.1]; p. 166.

Bender, Edward A., *Asymptotic methods in enumeration*, SIAM Review, 16 (1974) 485-515. [3.1.2]; p. 61.

Berge, Claude, *Graph theory*, Amer. Math. Monthly, 71 (1964) 471-481. [3.2.1]; p. 63.

———, *Perfect graphs*, in D.R. Fulkerson, Studies in Graph Theory, Part I, M.A.A., 1975, pp. 1-22. [3.2.3]; p. 65.

Bergmann, Peter G., *Fifty years of relativity*, Science, 123 (1956) 487-494. [1.2.1]; p. 6.

Berkovitz, Leonard D., *Optimal control theory*, Amer. Math. Monthly, 83 (1976) 225-239. [7.6.2]; p. 211.

Bernays, Paul, *On Platonism in mathematics*, in P. Benacerraf and H. Putnam, Philosophy of Mathematics, Prentice-Hall, 1964, pp. 274-286. [2.2.1]; p. 44.

Bernhard, Robert, *Crisis in math--is there 'universal truth'?*, Scientific Research, (October 14, 1968) 47-56. [2.5.0]; p. 52.

———, *Heresy in the halls of biology: mathematicians question Darwinism*, Scientific Research, (November 1967) 59-66. [7.2.0]; p. 196.

Bernhart, Arthur, *Curves of pursuit*, Scripta Math., 20 (1954) 125-141; 23 (1957) 49-65. [4.5.2]; p. 109.

———, *Polygons of pursuit*, Scripta Math., 24 (1959) 23-50. [4.5.2]; p. 109.

———, *Curves of general pursuit*, Scripta Math., 24 (1959) 189-206. [4.5.2]; p. 109.

Bernkopf, Michael, *A history of infinite matrices*, Arch. Hist. Exact Sci., 4 (1968) 308-358. [1.2.3]; p. 12.

———, *The development of function spaces with particular reference to their origins in integral equation theory*, Arch.Hist. Exact Sci., 3 (1966) 1-96. [1.2.3]; p. 12.

Bernstein, Allan R., *Non-standard analysis*, in M.D. Morley, Studies in Model Theory, M.A.A., 1973, pp. 35-58. [2.4.3]; p. 50.

Bers, Lipman, *Complex analysis*, in The Mathematical Sciences--A Collection of Essays for COSRIMS, M.I.T., 1969, pp. 7-20. [4.4.1]; p. 104.

———, *Uniformization moduli and Kleinian groups*, Bull. London Math. Soc., 4 (1972) 257-300. [4.4.3]; p. 107.

Bhat, U. Narayan, *Sixty years of queueing theory*, Management Science, 15:6 (1969) B280-B294. [6.1.2]; p. 157.

Billingsley, Patrick, *On the central limit theorem for the prime divisor function*, Amer. Math. Monthly, 76 (1969) 132-139. [3.3.2]; p. 69.

———, *Prime numbers and Brownian motion*, Amer. Math. Monthly, 80 (1973) 1099-1115. [3.3.2]; p. 70.

Binder, Raymond C., et al., *Mathematical aspects of physical theories*, Encyclopaedia Britannica, 15th ed., 1974, Macropaedia V. 14, pp. 392-424. [7.1.2]; p. 186.

Bing, R.H., *Challenging conjectures*, Amer. Math. Monthly, 74 (1967) Suppl. pp. 56-64. [5.5.2]; p. 141.

———, *Elementary point set topology*, Amer. Math. Monthly, 67 (1960) Suppl. pp. 1-58. [5.5.2]; p. 141.

———, *General topology*, Encyclopaedia Britannica, 15th ed., 1974, Macropaedia V. 18, pp. 509-514. [5.5.2]; p. 141.

———, *Point-set topology*, in The Mathematical Sciences--A Collection of Essays for COSRIMS, M.I.T., 1969, pp. 209-216. [5.5.0]; p. 139.

———, *Some aspects of the topology of 3-manifolds related to the Poincaré conjecture*, in T.L. Saaty, Lectures on Modern Mathematics, V. 2, Wiley, 1964, pp. 93-128. [5.6.2]; p. 145.

———, *Spheres in E-3*, Amer. Math. Monthly, 71 (1964) 353-364. [5.5.2]; p. 141.

———, *The elusive fixed point property*, Amer. Math. Monthly, 76 (1969) 119-132. [5.5.3]; p. 143.

Birkhoff, Garrett, *Boolean algebra*, McGraw-Hill Encyclopedia of Science and Technology, 1960, V. 2, pp. 288-290. [3.6.1]; p. 84.

———, *Current trends in algebra*, Amer. Math. Monthly, 80 (1973) 760-782; Correction, 81 (1974) 746. [1.2.2]; p. 10.

———, *Mathematics and computer science*, Amer. Scientist, 63 (1975) 83-91. [6.5.0]; p. 177.

———, *Mathematics and psychology*, SIAM Review, 11 (1969) 429-469. [7.3.1]; p. 201.

———, et al., *Algebraic structures*, Encyclopaedia Britannica, 15th ed., 1974, Macropaedia V. 1, pp. 518-558. [3.6.3]; p. 86.

Birkhoff, George D., Aesthetic Measure, Harvard U. Pr., 1933. [1.7.0]; p. 36.

———, *The mathematical nature of physical theories*, Amer. Scientist, 31 (1943) 281-310. [1.4.0]; p. 21.

———, *What is the ergodic theorem?*, Amer. Math. Monthly, 49 (1942) 222-226. [4.7.2]; p. 121.

Birnbaum, Allan, *On the foundations of statistical inference*, J. Amer. Stat. Assoc., 57 (1962) 269-326. [6.2.3]; p. 163.

Black, Max, *Induction*, Encyclopedia Americana, 1976, V. 15, p. 100. [2.6.1]; p. 55.

———, *The elusiveness of sets*, Review of Metaphysics, 24 (1971) 614-636. [2.1.1]; p. 39.

Bliss, Gilbert A., *Algebraic functions and their divisors*, Annals of Math., 26 (1924-25) 95-124. [3.6.3]; p. 87.

Blumenthal, Leonard M., *"A paradox, a paradox, a most ingenious paradox,"* Amer. Math. Monthly, 47 (1940) 346-353. [4.3.1]; p. 99.

———, *Analytic geometry*, McGraw-Hill Encyclopedia of Science and Technology, 1960, V. 1, pp. 387-392. [5.1.0]; p. 125.

———, *Logic*, McGraw-Hill Encyclopedia of Science and Technology, 1960, V. 7, pp. 578-582. [2.6.1]; p. 55.

Boas, Ralph P., Jr., *Inequalities for the derivatives of polynomials*, Math. Magazine, 42 (1969) 165-174. [4.1.1]; p. 91.

———, *Inversion of Fourier and Laplace transforms*, Amer. Math. Monthly, 69 (1962) 955-960. [4.2.2]; p. 96.

Bochner, Salomon, *Mathematical reflections*, Amer. Math. Monthly, 81 (1974) 827-852. [1.2.1]; p. 6.

———, *Mathematics*, McGraw-Hill Encyclopedia of Science and Technology, 1960, V. 8, pp. 175-180. [1.1.1]; p. 2.

———, *The rise of functions*, Rice Univ. Studies, 56:2 (1970) 3-21. [1.2.2]; p. 10.

———, *The role of mathematics in the rise of mechanics*, Amer. Scientist, 50 (1962) 294-311. [1.2.1]; p. 7.

Boehm, George A.W., <u>The New World of Mathematics</u>, The Dial Press, 1959. [1.1.0]; p. 1.

Bohnenblust, H. Frederic, *The theory of games*, in E.F. Beckenbach, <u>Modern Mathematics for the Engineer</u>, McGraw-Hill, 1956, pp. 191-210. [6.3.1]; p. 166.

Bolker, Ethan D., *The spinor spanner*, <u>Amer. Math. Monthly</u>, 80 (1973) 977-984. [7.1.2]; p. 187.

Bonsall, F.F., *A survey of Banach algebra theory*, <u>Bull. London Math. Soc.</u>, 2 (1970) 257-274. [4.6.3]; p. 117.

Boone, William W., *The word problem*, <u>Annals of Math.</u>, 70 (1959) 207-265. [3.5.3]; p. 81.

Bos, H.J.M., *Differentials, higher-order differentials and the derivative in the Leibnizian calculus*, <u>Arch. Hist. Exact Sci.</u>, 14 (1974) 1-90. [1.2.1]; p. 7.

Bose, Raj C. and Grünbaum, Branko, *Combinatorics and combinatorial geometry*, <u>Encyclopaedia Britannica</u>, 15th ed., 1974, Macropaedia V. 4, pp. 942-954. [5.2.2]; p. 131.

Bott, Raoul, *On the shape of a curve*, <u>Advances in Math.</u>, 16 (1975) 144-159. [5.4.2]; p. 137.

────, *The periodicity theorem for the classical groups and some of its applications*, <u>Advances in Math.</u>, 4 (1970) 353-411. [5.7.3]; p. 149.

Botts, Truman, *Probability theory and the Lebesgue integral*, <u>Math. Magazine</u>, 42 (1969) 105-111. [4.3.2]; p. 99.

Bourbaki, Nicolas, *Foundations of mathematics for the working mathematician*, <u>J. Symbolic Logic</u>, 14 (1949) 1-8. [2.2.2]; p. 45.

────, *The architecture of mathematics*, <u>Amer. Math. Monthly</u>, 57 (1950) 221-232, in F. LeLionnais, <u>Great Currents of Mathematical Thought</u>, V. 1, Dover, 1971, pp. 23-36. [1.5.0]; p. 25.

Boudreau, P.E., Griffin, J.S., Jr. and Kac, Mark, *An elementary queueing problem*, <u>Amer. Math. Monthly</u>, 69 (1962) 713-724. [6.1.2]; p. 157.

Boyd, John Paul, *The algebra of group kinship*, <u>J. Math. Psych.</u>, 6 (1969) 139-167. [7.3.2]; p. 202.

Boyer, Lee E., *Arithmetic*, <u>Encyclopedia Americana</u>, 1976, V. 2, pp. 293-297. [3.6.0]; p. 84.

Brainerd, Charles J., *The origins of number concepts*, <u>Scientific American</u>, 228 (March 1973) 100-109, 128. [1.6.0]; p. 32.

Brand, Louis, *A division algebra for sequences and its associated operational calculus*, <u>Amer. Math. Monthly</u>, 71 (1964) 719-728. [4.2.2]; p. 96.

────, *The pi theorem of dimensional analysis*, <u>Arch. Rat. Mech. Anal.</u>, 1 (1957) 34-45. [7.1.0]; p. 184.

Brauer, Alfred, *Matrix*, <u>Encyclopedia Americana</u>, 1976, V. 18, pp. 437-438. [3.4.1]; p. 75.

Brauer, Richard, *Representations of finite groups*, in T.L. Saaty, Lectures on Modern Mathematics, V. 1, Wiley, 1963, pp. 133-175. [3.5.3]; p. 81.

Brearley, M.N., *The long jump miracle of Mexico City*, Math. Magazine, 45 (1972) 241-246. [7.2.1]; p. 196.

Bremermann, H.J., *Several complex variables*, in I.I. Hirschman, Jr., Studies in Real and Complex Analysis, M.A.A., 1965, pp. 3-33. [4.4.3]; p. 107.

Brennan, Jean F., *Exploiting sparse matrices*, IBM Research Reports, 6:1 (1970) 1-8. [6.4.0]; p. 172.

———, *The fastest time of addition and multiplication*, IBM Research Reports, 4:1 (1968) 1-8. [6.5.1]; p. 179.

Brockett, R.W., *The synthesis of dynamical systems*, Quarterly of Applied Math., 30 (1972) 41-50. [6.4.2]; p. 173.

Bronowski, Jacob, *The logic of the mind*, Amer. Scientist, 54 (1966) 1-14. [1.4.0]; p. 21.

———, *The music of the spheres*, in J. Bronowski, The Ascent of Man, Little, Brown, 1973, pp. 154-187. [1.4.0]; p. 21.

Brouwer, L.E.J., *Historical background, principles, and methods of intuitionism*, South African J. Sci., 49 (1952-53) 139-146. [2.3.1]; p. 47.

———, *Intuitionism and formalism*, Bull. Amer. Math. Soc., 20 (1913) 81-96; also in P. Benacerraf and H. Putnam, Philosophy of Mathematics, Prentice-Hall, 1964, pp. 66-77. [2.3.1]; p. 47.

Browder, Felix E., *Is mathematics relevant? And if so, to what?*, Univ. of Chicago Magazine, 67:3 (Spring 1975) 11-16; also appears as *The relevance of mathematics*, Amer. Math. Monthly, 83 (1976) 249-254. [1.4.0]; p. 21.

Brown, Brenda W. and Brown, Robert F., *Hardy's "Apology": classic essay or cloistral clowning?*, Delta, 3:3 (Spring 1973) 1-10. [1.5.0]; p. 25.

Brown, Robert F., *Elementary consequences of the noncontractibility of the circle*, Amer. Math. Monthly, 81 (1974) 247-252. [5.5.2]; p. 141.

——— and Brown, Brenda W., *Hardy's "Apology": classic essay or cloistral clowning?*, Delta, 3:3 (Spring 1973) 1-10. [1.5.0]; p. 25.

Brown, Thomas A. and Roberts, Fred S., *Signed digraphs and the energy crisis*, Amer. Math. Monthly, 82 (1975) 577-594. [7.6.1]; p. 211.

Brualdi, R.A., *Transversal theory and graphs*, in D.R. Fulkerson, Studies in Graph Theory, Part I, M.A.A., 1975, pp. 23-88. [3.2.2]; p. 64.

Bruck, R.H., *Recent advances in the foundations of Euclidean plane geometry*, Amer. Math. Monthly, 62 (1955) Suppl. pp. 2-17. [5.1.2]; p. 128.

———, *What is a loop?*, in A.A. Albert, Studies in Modern Algebra, M.A.A., 1963, pp. 59-99. [3.6.2]; p. 85.

Bruckner, Andrew M., *Derivatives: why they elude classification*, Math. Magazine, 49 (1976) 5-11. [4.2.2]; p. 96.

———, *Differentiation of integrals*, Amer. Math. Monthly, 78 (1971) Suppl. pp. 1-51. [4.3.3]; p. 102.

——— and Leonard, J.L., *Derivatives*, Amer. Math. Monthly, 73 (1966) Suppl. pp. 24-56. [4.2.2]; p. 96.

Brun, Viggo, *Euclidean algorithms and musical theory*, L'Enseignement Math., 10 (1964) 125-137. [7.6.0]; p. 210.

Brush, Stephen G., *Foundations of statistical mechanics 1845-1915*, Arch. Hist. Exact Sci., 4 (1967) 145-183. [7.1.2]; p. 187.

Buck, R.C., *Mathematical induction and recursive definitions*, Amer. Math. Monthly, 70 (1963) 128-135. [2.6.1]; p. 55.

———, *Topology and analysis*, Math. Magazine, 40 (1967) 71-74. [4.2.2]; p. 96.

Budden, F.J., *Modern mathematics and music*, Math. Gazette, 51 (1967) 204-215. [7.6.0]; p. 210.

Busemann, Herbert, *Non-Euclidean geometry*, Math. Magazine, 24 (1950) 19-34. [5.1.1]; p. 126.

Butchart, J.H. and Moser, Leo, *No calculus, please*, Scripta Math., 18 (1952) 221-236. [4.1.1]; p. 91.

Butson, Alton T. and Andrus, Jan F., *Ordered groups*, Amer. Math. Monthly, 70 (1963) 619-628. [3.5.2]; p. 80.

Callahan, James, *Singularities and plane maps*, Amer. Math. Monthly, 81 (1974) 211-240. [5.6.2]; p. 145.

Cameron, R.H., *Some introductory exercises in the manipulation of Fourier transforms*, National Math. Magazine, 15 (1941) 331-356. [4.7.3]; p. 122.

Cannon, John J., *Computers in group theory: a survey*, Comm. Assoc. Comp. Mach., 12 (1969) 3-12. [3.5.3]; p. 81.

Carlson, Frederic R., Jr. and Zobrist, Albert L., *An advice-taking chess computer*, Scientific American, 228 (June 1973) 92-105, 124. [6.5.0]; p. 178.

Carnap, Rudolf, *What is probability?*, Scientific American, 189 (September 1953) 128-138, 170; also in D. Messick, Mathematical Thinking in Behavioral Sciences, Freeman, 1968, pp. 14-17, 223. [6.1.0]; p. 154.

Carrell, James B. and Dieudonné, Jean A., *Invariant theory, old and new*, in Advances in Math., 4 (1969) 1-80. [3.6.3]; p. 87.

Carrier, George F., *Singular perturbation theory and geophysics*, SIAM Review, 12 (1970) 175-193; also in A.H. Taub, Studies in Applied Mathematics, M.A.A., 1971, pp. 1-26. [7.1.3]; p. 190.

Carter, F.L., *Perspective drawing by numbers*, Math. Gazette, 53 (1969) 133-139. [7.6.1]; p. 211.

Carter, R.W., *Simple groups and simple Lie algebras*, J. London Math. Soc., 40 (1965) 193-240. [3.5.3]; p. 82.

Cartwright, Mary L., *Mathematics and thinking mathematically*, Amer. Math. Monthly, 77 (1970) 20-28. [1.5.0]; p. 25.

――――, *The mathematical mind*, Math. Spectrum, 2 (1969-70) 37-45. [1.5.0]; p. 25.

Casari, Ettore, *Axiomatical and set-theoretical thinking*, Synthese, 27 (1974) 49-61. [2.2.2]; p. 45.

Cassels, J.W.S., *Diophantine equations with special reference to elliptic curves*, J. London Math. Soc., 41 (1966) 193-291. [3.3.3]; p. 72.

Casten, R.G. and Lagerstrom, P.A., *Basic concepts underlying singular perturbation techniques*, SIAM Review, 14 (1972) 63-120. [7.1.3]; p. 190.

Cauman, Leigh S., *On indirect proof*, Scripta Math., 28 (1968) 101-115. [2.2.2]; p. 45.

Cesari, Lamberto, *Recent results in surface area theory*, Amer. Math. Monthly, 66 (1959) 173-192. [4.2.2]; p. 96.

――――, *Surface area*, in S.S. Chern, Studies in Global Geometry and Analysis, M.A.A., 1967, pp. 123-146. [4.2.2]; p. 96.

Chaitin, Gregory J., *Randomness and mathematical proof*, Scientific American, 232 (May 1975) 47-52, 122. [2.5.0]; p. 52.

Chakerian, G.D., *Intersection and covering properties of convex sets*, Amer. Math. Monthly, 76 (1969) 753-766. [5.2.2]; p. 131.

―――― and Lange, L.H., *Geometric extremum problems*, Math. Magazine, 44 (1971) 57-69. [6.3.1]; p. 166.

Chandler, Bruce, *Field*, Encyclopedia Americana, 1976, V. 11, pp. 163-164. [3.6.1]; p. 84.

――――, *Group*, Encyclopedia Americana, 1976, V. 13, pp. 515-516. [3.5.1]; p. 79.

Chandrasekhar, S., *Characteristic-value problems in hydrodynamic and hydromagnetic theory*, in E.F. Beckenbach, Modern Mathematics for the Engineer, 2nd Ser., McGraw-Hill, 1961, pp. 338-346. [7.1.2]; p. 187.

Chang, C.C., *What's so special about saturated models?*, in M.D. Morley, Studies in Model Theory, M.A.A., 1973, pp. 59-95. [2.6.3]; p. 57.

Chen, Wai-Kai, *Boolean matrices and switching nets*, Math. Magazine, 39 (1966) 1-8. [3.6.1]; p. 84.

Chern, S.S., *Curves and surfaces in Euclidean space*, in S.S. Chern, Studies in Global Geometry and Analysis, M.A.A., 1967, pp. 16-56. [5.3.3]; p. 134.

―――, *Differential geometry*, Encyclopaedia Britannica, 15th ed., 1974, Macropaedia V. 7, pp. 1093-1099. [5.3.3]; p. 134.

―――, *Differential geometry: its past and its future*, in Actes Cong. Inter. Math. (1970), V. 1, Gauthier-Villars, 1971, pp. 41-53. [5.3.3]; p. 135.

Chinn, William G. and Steenrod, Norman E., First Concepts of Topology, New Math. Libr., No. 18, Random House, 1966; M.A.A., 1975. [5.5.1]; p. 140.

Chittendon, J. Brace, *Quadrature of the circle*, Encyclopedia Americana, 1976, V. 23, pp. 52-53. [1.2.1]; p. 7.

Christie, Dan E., *Some thermodynamic properties of a mapping*, J. Franklin Inst., 267 (1959) 119-133. [7.1.2]; p. 187.

Chunikhin, S.A., *Some trends in the development of the theory of finite groups in recent years*, Russian Math. Surveys, 16:4 (1961) 29-46. [3.5.3]; p. 82.

Clark, Colin W., *Profit maximization and the extinction of animal species*, J. Political Economy, 81 (1973) 950-961. [7.2.1]; p. 196.

―――, *Some socially relevant applications of elementary calculus*, Two-Year College Mathematics Journal, 4 (1973) 1-15. [4.1.1]; p. 91.

―――, *The dynamics of commercially exploited natural animal populations*, Math. Biosciences, 13 (1972) 149-164. [7.2.1]; p. 196.

―――, *The economics of overexploitation*, Science, 181 (1973) 630-634. [7.4.1]; p. 204.

Clark, R.M., *Statistics and radiocarbon dating*, Math. Spectrum, 7 (1974-75) 83-89. [7.6.0]; p. 210.

Clements, Forrest E., *Use of cluster analysis with anthropological data*, Amer. Anthropologist, 56 (1954) 180-199. [7.3.1]; p. 201.

Cohen, Hirsh, *Mathematical applications, computation, and complexity*, Quarterly of Applied Math., 30 (1972) 109-121. [1.4.1]; p. 22.

―――, *Mathematics and the biomedical sciences*, in The Mathematical Sciences--A Collection of Essays for COSRIMS, M.I.T., 1969, pp. 217-231. [7.2.1]; p. 197.

―――, *Nonlinear diffusion problems*, in A.H. Taub, Studies in Applied Mathematics, M.A.A., 1971, pp. 27-64. [7.1.3]; p. 190.

Cohen, Joel E., *Information theory and music*, Behavioral Science, 7 (1962) 137-163. [7.6.0]; p. 210.

Cohen, John, *Subjective probability*, Scientific American, 197 (November 1957) 128-138, 184; also in D. Messick, Mathematical Thinking in Behavioral Sciences, Freeman, 1968, pp. 18-22, 223. [6.1.0]; p. 154.

Cohen, Maurice, *Foliations of 3-manifolds*, Amer. Math. Monthly, 81 (1974) 462-473. [5.6.3]; p. 146.

Cohen, Paul J., *Comments on the foundations of set theory*, in D. Scott, Axiomatic Set Theory, Part 1, Amer. Math. Soc., 1971, pp. 9-15. [2.1.2]; p. 39.

——— and Hersh, Reuben, *Non-Cantorian set theory*, Scientific American, 217 (December 1967) 104-116, 160; also in M. Kline, Mathematics in the Modern World, Freeman, 1968, pp. 212-220, 400. [2.1.1]; p. 39.

Cohn, P.M., *Algebra and language theory*, Bull. London Math. Soc., 7 (1975) 1-29. [3.6.3]; p. 87.

———, *Free associative algebras*, Bull. London Math. Soc., 1 (1969) 1-39. [3.6.3]; p. 87.

———, *Rings of fractions*, Amer. Math. Monthly, 78 (1971) 596-615. [3.6.2]; p. 85.

———, *Unique factorization domains*, Amer. Math. Monthly, 80 (1973) 1-18; Correction, 80 (1973) 1115. [3.6.2]; p. 85.

Cohn-Vossen, Stephan and Hilbert, David, Geometry and the Imagination, Chelsea, 1952. [5.5.1]; p. 140.

Coifman, R.R. and Weiss, Guido, *Representations of compact groups and spherical harmonics*, L'Enseignement Math., 14 (1968) 121-173. [4.7.3]; p. 122.

Collins, G.E., *Computer algebra of polynomials and rational functions*, Amer. Math. Monthly, 80 (1973) 725-755. [6.5.2]; p. 180.

Connel, E.H., *A classical theorem in complex variables*, Amer. Math. Monthly, 72 (1965) 729-732. [4.4.2]; p. 105.

Cooley, H.R., *Algebra*, McGraw-Hill Encyclopedia of Science and Technology, 1960, V. 1, pp. 238-241. [3.6.0]; p. 84.

Coolidge, J.L., *The lengths of curves*, Amer. Math. Monthly, 60 (1953) 89-93. [1.2.1]; p. 7.

———, *The number e*, Amer. Math. Monthly, 57 (1950) 591-602; also in T.M. Apostol, Selected Papers on Calculus, M.A.A., 1969, pp. 8-19. [4.1.1]; p. 91.

Coppel, W.A., *J.B. Fourier--On the occasion of his two hundredth birthday*, Amer. Math. Monthly, 76 (1969) 468-483. [1.2.3]; p. 12.

Costello, W.G. and Taylor, H.M., *Deterministic population growth models*, Amer. Math. Monthly, 78 (1971) 841-855. [7.2.2]; p. 198.

Courant, Richard, *Gauss and the present situation of the exact sciences*, in T.L. Saaty and F.J. Weyl, The Spirit and the Uses of the Mathematical Sciences, McGraw-Hill, 1969, pp. 141-155. [1.5.0]; p. 25.

———, *Mathematics in the modern world*, Scientific American, 211 (September 1964) 40-49, 269; also in M. Kline, Mathematics in the Modern World, Freeman, 1968, pp. 19-27, 394. [1.4.1]; p. 22.

———, *Soap film experiments with minimal surfaces*, Amer. Math. Monthly, 47 (1940) 167-174. [5.3.1]; p. 133.

────── and Robbins, Herbert, *Topology*, in R. Courant and H. Robbins, What is Mathematics, Oxford U. Pr., 1941, pp. 235-271; also in J.R. Newman, The World of Mathematics, V. 1, Simon and Schuster, 1956, pp. 581-599. [5.5.0]; p. 139.

────── and ──────, *What is Mathematics?*, Oxford U. Pr., 1941. [1.1.1]; p. 2.

Cowan, Thaddeus M., *The theory of braids and the analysis of impossible figures*, J. Math. Psych., 11 (1974) 190-212. [7.6.2]; p. 211.

Coxeter, H.S. MacDonald, *A geometrical background for de Sitter's world*, Amer. Math. Monthly, 50 (1943) 217-228. [5.1.1]; p. 126.

──────, *Geometry*, in T.L. Saaty, Lectures on Modern Mathematics, V. 3, Wiley, 1965, pp. 58-94. [5.1.2]; p. 128.

──────, *Map coloring problems*, Scripta Math., 23 (1957) 11-25. [3.2.1]; p. 63.

──────, *Music and mathematics*, Math. Teacher, 61 (1968) 312-320; also in Canad. Music J., 6 (1962) pp. 13-24. [7.6.0]; p. 210.

──────, *Non-Euclidean geometry*, Encyclopaedia Britannica, 15th ed., 1974, Macropaedia V. 7, pp. 1112-1120. [5.1.1]; p. 126.

──────, *Non-Euclidean geometry*, in The Mathematical Sciences--A Collection of Essays for COSRIMS, M.I.T., 1969, pp. 52-59. [5.1.0]; p. 125.

──────, *The problem of Apollonius*, Amer. Math. Monthly, 75 (1968) 5-15. [5.1.1]; p. 126.

──────, *The space-time continuum*, Historia Math., 2 (1975) 289-298. [1.2.1]; p. 7.

────── and Ball, W.W. Rouse, Mathematical Recreations & Essays, Twelfth Edition, U. of Toronto Pr., 1974. [1.7.0]; p. 36.

────── and Greitzer, S.L., Geometry Revisited, New Math. Libr., No. 19, Random House, 1967; M.A.A., 1975. [5.1.1]; p. 126.

Craig, William, *Unification and abstraction in algebraic logic*, in A. Daigneault, Studies in Algebraic Logic, M.A.A., 1974, pp. 6-57. [2.6.3]; p. 57.

Crossley, John N., *Recursive equivalence*, Bull. London Math. Soc., 2 (1970) 129-151. [2.6.3]; p. 57.

──────, et al., *What is Mathematical Logic?*, Oxford U. Pr., 1972. [2.6.2]; p. 56.

Crowe, Donald W., *The geometry of African art I. Bakuba art*, J. of Geometry, 1 (1971) 169-182; *II. A catalog of Benin patterns*, Historia Math., 2 (1975) 253-271. [7.6.1]; p. 211.

Crowe, Michael J., *Ten "laws" concerning patterns of change in the history of mathematics*, Historia Math., 2 (1975) 161-166. [1.2.0]; p. 4.

Cunningham, Frederic, Jr., *Taking limits under the integral sign*, Math. Magazine, 40 (1967) 179-186. [4.2.1]; p. 95.

―――, *The Kakeya problem for simply connected and for star-shaped sets*, Amer. Math. Monthly, 78 (1971) 114-129. [6.3.2]; p. 168.

Curtis, Charles W., *The classical groups as a source of algebraic problems*, Amer. Math. Monthly, 74 (1967) Suppl. pp. 80-91. [3.5.3]; p. 82.

―――, *The four and eight square problem and division algebras*, in A.A. Albert, Studies in Modern Algebra, M.A.A., 1963, pp. 100-125. [3.6.3]; p. 87.

Curtis, Edward B., *Simplicial homotopy theory*, Advances in Math., 6 (1971) 107-209. [5.7.3]; p. 149.

Curtiss, J.H., *Faber polynomials and the Faber series*, Amer. Math. Monthly, 78 (1971) 577-596; Correction, 79 (1972) 363. [4.4.3]; p. 107.

Danby, John M.A., *Celestial mechanics*, Encyclopedia Americana, 1976, V. 6, pp. 125-127. [7.1.1]; p. 185.

Dantzig, George B., *Maximization of a linear function of variables subject to linear inequalities*, in T.C. Koopmans, Activity Analysis of Production and Allocation, Wiley, 1951, pp. 339-347. [6.3.1]; p. 166.

―――, *New Mathematical methods in the life sciences*, Amer. Math. Monthly, 71 (1964) 4-15. [7.2.1]; p. 197.

―――, *On the shortest route through a network*, Management Science, 6 (January 1960); also in D.R. Fulkerson, Studies in Graph Theory, Part I, M.A.A., 1975, pp. 89-93. [6.3.1]; p. 166.

Dantzig, Tobias, Number, The Language of Science, Doubleday Anchor, 1956. [1.1.0]; p. 1.

Darrow, Karl K., *Memorial to the classical statistics*, Bell System Tech. J., 22 (1943) 108-135. [7.1.2]; p. 187.

―――, *The new statistical mechanics*, Bell System Tech. J., 22 (1943) 362-392. [7.1.2]; p. 187.

Darst, R.B., *Some Cantor sets and Cantor functions*, Math. Magazine, 45 (1972) 2-7. [4.2.2]; p. 96.

Dauben, Joseph W., *Denumerability and dimension: the origins of Georg Cantor's theory of sets*, Rete, 2 (1974) 105-133. [1.2.2]; p. 10.

―――, *The invariance of dimension: problems in the early development of set theory and topology*, Historia Math., 2 (1975) 273-288. [1.2.2]; p. 10.

―――, *The trigonometric background to Georg Cantor's theory of sets*, Arch. Hist. Exact Sci., 7 (1971) 181-216. [1.2.2]; p. 10.

David, F.N., Games, Gods and Gambling, Hafner, 1962. [1.2.0]; p. 4.

Davis, Chandler, *Materialist mathematics*, in R.S. Cohen and M.W. Wartofsky, Boston Studies in the Philosophy of Science, Humanities Pr., 1974, V. 15. [2.2.0]; p. 43.

—— and Pogorzelski, H.A., *Contemporary mathematical notation*, McGraw-Hill Encyclopedia of Science and Technology, 1960, V. 8, pp. 172-174. [1.2.1]; p. 7.

Davis, Martin, *Hilbert's tenth problem is unsolvable*, Amer. Math. Monthly, 80 (1973) 233-269. [3.3.3]; p. 72.

—— and Hersh, Reuben, *Hilbert's 10th problem*, Scientific American, 229 (November 1973) 84-91, 136. [3.3.1]; p. 68.

—— and ——, *Nonstandard analysis*, Scientific American, 226 (June 1972) 78-86, 136. [2.4.1]; p. 49.

Davis, Philip J., *Fidelity in mathematical discourse: is one and one really two?*, Amer. Math. Monthly, 79 (1972) 252-263. [2.2.1]; p. 44.

——, *Leonard Euler's integral: a historical profile of the gamma function*, Amer. Math. Monthly, 66 (1959) 849-869. [1.2.2]; p. 10.

——, *Number*, Scientific American, 211 (September 1964) 50-59, 269; also in M. Kline, Mathematics in the Modern World, Freeman, 1968, pp. 89-97, 396. [3.6.0]; p. 84.

——, *Numerical analysis*, in The Mathematical Sciences—A Collection of Essays for COSRIMS, M.I.T., 1969, pp. 128-137. [6.4.0]; p. 172.

——, *The Lore of Large Numbers*, New Math. Libr., No. 6, Random House, 1961; M.A.A., 1975. [3.3.0]; p. 68.

Deavours, C.A., *The quaternion calculus*, Amer. Math. Monthly, 80 (1973) 995-1008. [4.4.2]; p. 105.

Debreu, Gerard, *Mathematical theory of economic equilibrium*, in Proc. Inter. Cong. Math. (1974), V. 1, Canad. Cong. Math., 1975, pp. 65-77. [7.4.2]; p. 205.

De Broglie, Louis, *The role of mathematics in the development of contemporary theoretical physics*, in F. LeLionnais, Great Currents of Mathematical Thought, V. 2, Dover, 1971, pp. 78-93. [1.4.0]; p. 21.

Dehn, Max, *Mathematics*, Amer. Math. Monthly; *400 B.C.-300 B.C.*, 50 (1943) 411-414; *300 B.C.-200 B.C.*, 51 (1944) 25-31; *200 B.C.-600 A.D.*, 51 (1944) 149-157. [1.2.0]; p. 4.

De Jager, E.M., *Theory of distributions*, in E. Roubine, Mathematics Applied to Physics, Springer-Verlag, 1970, pp. 52-110. [4.6.3]; p. 118.

Delone, B.N., *Algebra: theory of algebraic equations*, in A.D. Aleksandrov, et al., Mathematics—Its Content, Methods and Meaning, V. 1, M.I.T., 1963, pp. 261-310. [3.6.1]; p. 84.

DeLong, Howard, *Unsolved problems in arithmetic*, Scientific American, 224 (March 1971) 50-60, 124. [2.5.1]; p. 52.

De Morgan, Augustus, A Budget of Paradoxes, Open Court, 1872; 1915; excerpted in J.R. Newman, The World of Mathematics, V. 4, Simon and Schuster, 1956, pp. 2369-2382. [1.7.0]; p. 36.

Desanti, Jean T., *From Cauchy to Riemann, or the birth of the theory of real functions*, in F. LeLionnais, Great Currents of Mathematical Thought, V. 1, Dover, 1971, pp. 181-190. [1.2.1]; p. 7.

DeSua, Frank C., *Consistency and completeness--a résumé*, Amer. Math. Monthly, 63 (1956) 295-305. [2.5.1]; p. 53.

——, *Metamathematics: a non-technical exposition*, Amer. Scientist, 42 (1954) 488-495. [2.5.0]; p. 52.

Diamond, Harold G. and Bateman, Paul T., *Asymptotic distribution of Beurling's generalized prime numbers*, in W.J. LeVeque, Studies in Number Theory, M.A.A., 1969, pp. 152-210. [3.3.3]; p. 72.

Diaz, J.B., *Differential equation*, McGraw-Hill Encyclopedia of Science and Technology, 1960, V. 4, pp. 125-128. [4.5.1]; p. 109.

Dickson, L.E., *The Waring problem and its generalizations*, Bull. Amer. Math. Soc., 42 (1936) 833-842. [3.3.1]; p. 68.

Dieudonné, Jean A., *Algebraic geometry*, Advances in Math., 3 (1969) 233-321. [5.4.3]; p. 137.

——, *Mathematics*, Collier's Encyclopedia, 1976, V. 15, pp. 541-552. [1.5.0]; p. 25.

——, *Recent developments in mathematics*, Amer. Math. Monthly, 71 (1964) 239-248. [1.1.2]; p. 3.

——, *Should we teach "modern" mathematics?*, Amer. Scientist, 61 (1973) 16-19. [1.6.0]; p. 32.

——, *The historical development of algebraic geometry*, Amer. Math. Monthly, 79 (1972) 827-866. [1.2.3]; p. 12.

——, *The work of Nicolas Bourbaki*, Amer. Math. Monthly, 77 (1970) 134-145. [1.3.1]; p. 19.

—— and Carrell, James B., *Invariant theory, old and new*, in Advances in Math., 4 (1969) 1-80. [3.6.3]; p. 87.

Doob, J.L., *Probability in function space*, Bull. Amer. Math. Soc., 53 (1947) 15-30. [4.3.2]; p. 100.

——, *What is a martingale?*, Amer. Math. Monthly, 78 (1971) 451-463. [6.1.2]; p. 157.

——, *What is a stochastic process?*, Amer. Math. Monthly, 49 (1942) 648-653. [6.1.1]; p. 156.

Douglas, Jesse, *The problem of Plateau*, Scripta Math., 5 (1938) 159-164. [5.3.1]; p. 133.

Drake, Stillman, *Galileo's discovery of the law of free fall*, Scientific American, 228 (May 1973) 84-92, 120. [1.2.1]; p. 7.

——, *Mathematics and discovery in Galileo's physics*, Historia Math., 1 (1974) 129-150. [1.2.1]; p. 8.

—— and MacLachlan, James, *Galileo's discovery of the parabolic trajectory*, Scientific American, 232 (March 1975) 102-110, 132. [7.1.0]; p. 184.

Dubins, L.E. and Spanier E.H., *How to cut a cake fairly*, Amer. Math. Monthly, 68 (1961) 1-17. [6.3.3]; p. 169.

Dudley, Patricia L., et al., *Further techniques in the theory of big game hunting*, Amer. Math. Monthly, 75 (1968) 896-897. [1.7.2]; p. 37.

Dudley, Underwood, *Who was the first non-Euclidean?*, Math. Spectrum, 6 (1973-74) 41-46. [5.1.0]; p. 125.

Duff, G.F.D., *Mathematical problems of tidal energy*, in Proc. Inter. Cong. Math. (1974), V. 1, Canad. Cong. Math., 1975, pp. 87-94. [7.1.2]; p. 187.

Duffin, R.J., *Network models*, SIAM-AMS Proc., 3 (1971) 65-91; an expanded version appears as *Electrical network models* in D.R. Fulkerson, Studies in Graph Theory, Part I, M.A.A., 1975, pp. 94-138. [7.1.2]; p. 187.

Du Val, Patrick, *Projective geometry*, Encyclopaedia Britannica, 15th ed., 1974, Macropaedia, V. 7, pp. 1120-1125. [5.1.1]; p. 127.

Dyer, Eldon, *The functors of algebraic topology*, in P.J. Hilton, Studies in Modern Topology, M.A.A., 1968, pp. 134-164. [5.7.3]; p. 149.

Dyson, Freeman J., *Mathematics in the physical sciences*, Scientific American, 211 (September 1964) 128-146, 269-270; also in M. Kline, Mathematics in the Modern World, Freeman, 1968, pp. 249-257, 401; and in The Mathematical Sciences--A Collection of Essays for COSRIMS, M.I.T., 1969, pp. 97-115. [7.1.0]; p. 184.

———, *Missed opportunities*, Bull. Amer. Math. Soc., 78 (1972) 635-652. [7.1.2]; p. 188.

Earl, James M. and Salkind, Charles T., The M.A.A. Problem Book III, New Math. Libr., No. 25, Random House, 1973; M.A.A., 1975. [1.7.1]; p. 37.

Eberlein, W.F., *The elementary transcendental functions*, Amer. Math. Monthly, 61 (1954) 386-392; also in T.M. Apostol, Selected Papers on Calculus, M.A.A., 1969, pp. 126-133. [4.1.1]; p. 91.

Eddington, Sir Arthur Stanley, *The theory of groups*, in J.R. Newman, The World of Mathematics, V. 3, Simon and Schuster, 1956, pp. 1558-1573. [3.5.0]; p. 79.

Edmunds, D.E., *Quasilinear second order elliptic and parabolic equations*, Bull. London Math. Soc., 2 (1970) 5-28. [4.5.3]; p. 111.

Eggleston, H.G., *The isoperimetric problem*, in N.J. Hardiman, Exploring University Mathematics, V. 1, Pergamon, 1967, pp. 95-120. [5.3.1]; p. 133.

Eilenberg, Samuel, *Algebraic topology*, in T.L. Saaty, Lectures on Modern Mathematics, V. 1, Wiley, 1963, pp. 98-114. [5.7.3]; p. 149.

———, *The algebraization of mathematics*, in The Mathematical Sciences--A Collection of Essays for COSRIMS, M.I.T., 1969, pp. 153-160. [1.5.0]; p. 25.

Einstein, Albert, *On the generalized theory of gravitation*, Scientific American, 182 (April 1950) 13-17, 72; also in M. Kline, Mathematics in the Modern World, Freeman, 1968, pp. 258-262, 401. [7.1.0]; p. 184.

Elder, Barbara, *Paths and knots as geometric groups*, Pentagon, 28 (1968) 3-15. [3.5.2]; p. 80.

Elementary and multivariate algebra, Encyclopaedia Britannica, 15th ed., 1974, Macropaedia V. 1, pp. 499-507. [3.6.1]; p. 84.

Elias, Peter, *The noisy channel coding theorem for erasure channels*, Amer. Math. Monthly, 81 (1974) 853-862. [7.5.3]; p. 209.

Ellison, W.J., *Waring's problem*, Amer. Math. Monthly, 78 (1971) 10-36. [3.3.2]; p. 70.

Emerson, John D., *Simple points of an affine algebraic variety*, Amer. Math. Monthly, 82 (1975) 132-147. [5.4.3]; p. 137.

Engel, Arthur, *The relevance of modern fields of applied mathematics for mathematical education*, Educ. Studies Math., 2 (1969-70) 257-269. [1.6.1]; p. 34.

Erdélyi, Arthur, *From delta functions to distributions*, in E.F. Beckenbach, Modern Mathematics for the Engineer, 2nd Ser., McGraw-Hill, 1961, pp. 5-50. [4.3.2]; p. 100.

Erdös, Paul, *Some recent advances and current problems in number theory*, in T.L. Saaty, Lectures on Modern Mathematics, V. 3, Wiley, 1965, pp. 196-244. [3.3.2]; p. 70.

Ershov, Y.L., et al., *Elementary theories*, Russian Math. Surveys, 20:4 (1965) 35-105. [2.5.2]; p. 53.

Evans, G.C., *Modern methods of analysis in potential theory*, Bull. Amer. Math. Soc., 43 (1937) 481-502. [4.4.3]; p. 107.

Evans, Trevor, *Nonassociative number theory*, Amer. Math. Monthly, 64 (1957) 299-309. [3.3.1]; p. 68.

Eves, Howard W., *Mathematics*, Encyclopedia Americana, 1976, V. 18, pp. 431-434. [1.1.0]; p. 1.

Faddeev, D.K., *Linear algebra*, in A.D. Aleksandrov, et al., Mathematics--Its Content, Methods and Meaning, V. 3, M.I.T., 1963, pp. 37-96. [3.4.1]; p. 75.

Fang, J., *Hilbert's problems*, Philosophia Mathematica, 6 (1969) 38-53. [1.2.1]; p. 8.

Feferman, Solomon and Kleene, Stephen C., *Foundations of mathematics*, Encyclopaedia Britannica, 15th ed., 1974, Macropaedia V. 11, pp. 630-639. [2.2.1]; p. 45.

Fefferman, Charles L., *Recent progress in classical Fourier analysis*, in Proc. Inter. Cong. Math. (1974), V. 1, Canad. Cong. Math., 1975, pp. 95-118. [4.7.3]; p. 122.

—— and Zygmund, Antoni, *Fourier analysis*, Encyclopaedia Britannica, 15th ed., 1974, Macropaedia V. 1, pp. 735-757. [4.7.2]; p. 122.

Fehr, Howard F., *Value and the study of mathematics*, Scripta Math., 21 (1955) 49-53. [1.4.0]; p. 21.

Feit, Walter, *The current situation in the theory of finite simple groups*, in Actes Cong. Inter. Math., (1970) V. 1, Gauthier-Villars, 1971, pp. 55-93. [3.5.3]; p. 82.

Fejes Toth, L., *What the bees know and what they do not know*, Bull. Amer. Math. Soc., 70 (1964) 468-481. [5.1.1]; p. 127.

Feller, William, *A direct proof of Stirling's formula*, Amer. Math. Monthly, 74 (1967) 1223-1225; Correction, 75 (1968) 518. [4.1.1]; p. 91.

——, *Chance processes and fluctuations*, in E.F. Beckenbach, Modern Mathematics for the Engineer, 2nd Ser., McGraw-Hill, 1961, pp. 167-181. [6.1.2]; p. 157.

——, *On Müntz' theorem and completely monotone functions*, Amer. Math. Monthly, 75 (1968) 342-350. [4.2.2]; p. 96.

——, *Probability*, McGraw-Hill Encyclopedia of Science and Technology, 1960, V. 10, pp. 624-630. [6.1.1]; p. 156.

——, *The fundamental limit theorems in probability*, Bull. Amer. Math. Soc., 51 (1945) 800-832. [6.1.2]; p. 157.

Ferguson, Rolfe P., *On Fermat's last theorem*, J. Undergraduate Math., 6 (1974) 1-14, 85-97; 7 (1975) 35-45. [3.3.2]; p. 70.

Ficken, F.A., *Mathematics and the layman*, Amer. Scientist, 52 (1964) 419-430. [1.5.0]; p. 26.

——, *Some uses of linear spaces in analysis*, Amer. Math. Monthly, 66 (1959) 259-275. [4.6.2]; p. 115.

Fillmore, P.A., *The shift operator*, Amer. Math. Monthly, 81 (1974) 717-723. [4.6.3]; p. 118.

Fink, A.M., *Almost periodic functions invented for specific purposes*, SIAM Review, 14 (1972) 572-581. [4.7.3]; p. 123.

Finkbeiner, Daniel T., *Vector and tensor analysis*, Encyclopaedia Britannica, 15th ed., 1974, Macropaedia V. 1, pp. 791-799. [3.6.2]; p. 86.

Fisher, Charles S., *Some social characteristics of mathematicians and their work*, Amer. J. Sociology, 78 (1973) 1094-1118. [1.2.0]; p. 4.

——, *The death of a mathematical theory: a study in the sociology of knowledge*, Arch. Hist. Exact Sci., 3 (1966) 137-159. [1.2.1]; p. 8.

Fisher, Irwin and Struik, Ruth R., *Nil algebras and periodic groups*, Amer. Math. Monthly, 75 (1968) 611-623. [3.5.3]; p. 82.

Flanders, Harley, *A proof of Minkowski's inequality for convex curves*, Amer. Math. Monthly, 75 (1968) 581-593. [6.3.2]; p. 168.

———, *Differential forms*, in S.S. Chern, Studies in Global Geometry and Analysis, M.A.A., 1967, pp. 57-95. [5.3.2]; p. 134.

———, *Methods of proof in linear algebra*, Amer. Math. Monthly, 63 (1956) 1-15. [3.4.3]; p. 76.

Flegg, Graham, From Geometry to Topology, Crane Russak, 1974. [5.5.1]; p. 140.

Fletcher, T.J., *Doing without calculus*, Math. Gazette, 55 (1971) 4-17. [4.1.1]; p. 91.

Ford, Lester R. and Murray, Francis J., *Mathematics as a calculatory science*, Encyclopaedia Britannica, 15th ed., 1974, Macropaedia, V. 11, pp. 671-696. [1.5.0]; p. 27.

Forder, Henry G. and Valentine, Frederick A., *Euclidean geometry*, Encyclopaedia Britannica, 15th ed., 1974, Macropaedia, V. 7, pp. 1099-1112. [5.1.1]; p. 127.

Forsythe, George E., *Pitfalls in computation, or why a math book isn't enough*, Amer. Math. Monthly, 77 (1970) 931-956. [6.4.1]; p. 173.

———, *Solving a quadratic equation on a computer*, in The Mathematical Sciences—A Collection of Essays for COSRIMS, M.I.T., 1969, pp. 138-152. [6.4.0]; p. 172.

———, *What to do till the computer scientist comes*, Amer. Math. Monthly, 75 (1968) 454-462. [6.5.0]; p. 177.

Foulis, D.J. and Randall, C.H., *An approach to empirical logic*, Amer. Math. Monthly, 77 (1970) 363-374. [2.6.1]; p. 55.

Fraenkel, Abraham A., *On the crisis of the principle of the excluded middle*, Scripta Math., 17 (1951) 5-16. [2.3.0]; p. 47.

———, *The recent controversies about the foundations of mathematics*, Scripta Math., 13 (1947) 17-36. [2.2.2]; p. 45.

Frankel, Theodore, *Maxwell's equations*, Amer. Math. Monthly, 81 (1974) 343-349. [7.1.2]; p. 188.

Franklin, Philip, *Series*, McGraw-Hill Encyclopedia of Science and Technology, 1960, V. 12, pp. 189-196. [4.2.1]; p. 95.

———, *The four color problem*, Scripta Math., 6 (1939) 149-156, 197-210; excerpted in Galois Lectures, Scripta Mathematica, 1941, pp. 53-85. [3.2.0]; p. 63.

———, *What is topology?*, Phil. Sci., 2 (1935) 39-47. [5.5.0]; p. 139.

Freudenthal, Hans, *Lie groups in the foundations of geometry*, Advances in Math., 1 (1965) 145-190. [5.8.3]; p. 152.

———, Mathematics as an Educational Task, D. Reidel, 1973. [1.6.1]; p. 34.

———, *Models in applied probability*, Synthese, 12 (1960) 202-212. [6.1.1]; p. 156.

Friedrichs, K.O., From Pythagoras to Einstein, New Math. Libr., No. 16, Random House, 1965; M.A.A., 1975. [7.1.1]; p. 185.

Fromm, J.E. and Harlow, F.H., Computer experiments in fluid dynamics, Scientific American, 212 (March 1965) 104-110, 140. [7.1.0]; p. 184.

Fulkerson, D.R., Flow networks and combinatorial operations research, Amer. Math. Monthly, 73 (1966) 115-138; also in D.R. Fulkerson, Studies in Graph Theory, Part I, M.A.A., 1975, pp. 139-171. [6.3.1]; p. 166.

Gale, David, A mathematical theory of optimal economic development, Bull. Amer. Math. Soc., 74 (1968) 207-223. [7.4.2]; p. 205.

———, How to solve linear inequalities, Amer. Math. Monthly, 76 (1969) 589-599. [6.3.1]; p. 166.

———, Mathematics and economic models, Amer. Scientist, 44 (1956) 33-44. [7.4.1]; p. 204.

———, On the theory of interest, Amer. Math. Monthly, 80 (1973) 853-868. [7.4.2]; p. 205.

———, The jeep once more or jeeper by the dozen, Amer. Math. Monthly, 77 (1970) 493-501; Correction, 78 (1971) 644-645. [6.3.1]; p. 167.

——— and Shapley, L.S., College admissions and the stability of marriage, Amer. Math. Monthly, 69 (1962) 9-15. [6.3.0]; p. 165.

Gallager, Robert G., Information theory, Encyclopaedia Britannica, 15th ed., 1974, Macropaedia V. 9, pp. 574-580. [7.5.1]; p. 208.

Gans, David, An introduction to elliptic geometry, Amer. Math. Monthly, 62 (1955) Suppl. pp. 66-75. [5.1.1]; p. 127.

Gårding, Lars, Partial differential equations: problems and uniformization in Cauchy's problem, in T.L. Saaty, Lectures on Modern Mathematics, V. 2, Wiley, 1964, pp. 129-150. [4.5.3]; p. 111.

Gardner, Martin, Logic Machines, Diagrams and Boolean Algebras, Dover, 1968. [2.6.0]; p. 55.

———, Martin Gardner's Sixth Book of Mathematical Games from Scientific American, Freeman, 1971. [1.7.0]; p. 36.

———, Mathematical Carnival, Knopf, 1975. [1.7.0]; p. 36.

———, Mathematics, Magic and Mystery, Dover, 1956. [1.7.0]; p. 36.

———, New Mathematical Diversions from Scientific American, Simon and Schuster, 1966. [1.7.0]; p. 36.

———, Relativity for the Million, Macmillan, 1962. [7.1.0]; p. 184.

———, The Ambidextrous Universe, Basic Books, 1964. [7.1.0]; p. 184.

———, The Numerology of Dr. Matrix, Simon and Schuster, 1967. [1.7.0]; p. 36.

―――, *The Scientific American Book of Mathematical Puzzles and Diversions*, Simon and Schuster, 1959. [1.7.0]; p. 36.

―――, *The Second Scientific American Book of Mathematical Puzzles and Diversions*, Simon and Schuster, 1961. [1.7.0]; p. 36.

―――, *The Unexpected Hanging, and Other Mathematical Diversions*, Simon and Schuster, 1969. [1.7.0]; p. 36.

Gelbaum, B.R., *Banach algebras and their applications*, Amer. Math. Monthly, 71 (1964) 248-256. [4.6.2]; p. 115.

Gel'fand, I.M., *Functional analysis*, in A.D. Aleksandrov, et al., Mathematics--Its Content, Methods and Meaning, V. 3, M.I.T., 1963, pp. 227-261. [4.6.1]; p. 115.

Geoffrion, A.M. and Marsten, R.E., *Integer programming algorithms: a framework and state-of-the-art survey*, Management Science, 18 (1972) 465-491. [6.3.3]; p. 169.

Giblin, P.J., *What is an asymptote?*, Math. Gazette, 56 (1972) 274-284. [4.1.1]; p. 91.

Gilbert, E.N., *An outline of information theory*, Amer. Statistician, 12:1 (1958) 13-19. [7.5.0]; p. 208.

Glasser, M. Lawrence, *The summation of series*, SIAM J. Math. Anal., 2 (1971) 595-600. [6.4.2]; p. 173.

Gleason, Andrew M., *Evolution of an active mathematical theory*, Science, 145 (1964) 451-457; also appeared as *The evolution of differential topology*, in The Mathematical Sciences--A Collection of Essays for COSRIMS, M.I.T., 1969, pp. 176-189. [1.2.1]; p. 8.

Glushkov, V.M., *The abstract theory of automata*, Russian Math. Surveys, 16:5 (1961) 1-54. [3.6.3]; p. 87.

Gödel, Kurt, *What is Cantor's continuum problem?*, Amer. Math. Monthly, 54 (1947) 515-525; also in P. Benacerraf and H. Putnam, Philosophy of Mathematics, Prentice-Hall, 1964, pp. 258-273. [2.1.2]; p. 39.

Goffman, Casper, *Completeness of the real numbers*, Math. Magazine, 47 (1974) 1-8. [4.3.2]; p. 100.

―――, *Preliminaries to functional analysis*, in R.C. Buck, Studies in Modern Analysis, M.A.A., 1962, pp. 138-180. [4.6.2]; p. 115.

――― and Waterman, Daniel, *Some aspects of Fourier series*, Amer. Math. Monthly, 77 (1970) 119-133. [4.3.3]; p. 102.

Goldie, A.W., *Some aspects of ring theory*, Bull. London Math. Soc., 1 (1969) 129-154. [3.6.3]; p. 87.

Goldstein, L.J., *A history of the prime number theorem*, Amer. Math. Monthly, 80 (1973) 599-615; Correction, 80 (1973) 1115. [3.3.2]; p. 70.

Goldstein, Marie, *The historical development of group theoretical ideas in connection with Euclid's axiom of congruence*, Notre Dame J. of Formal Logic, 13 (1972) 331-349. [1.2.2]; p. 10.

G

Goldstine, H., The Computer from Pascal to von Neumann, Princeton U. Pr., 1972. [6.5.0]; p. 177.

Golomb, Solomon W. and Baumert, Leonard D., Backtrack programming, J. Assoc. Comp. Mach., 12 (1965) 516-524. [6.5.2]; p. 180.

Golos, Ellery B., Projective geometry, Encyclopedia Americana, 1976, V. 12, pp. 487-492. [5.1.1]; p. 127.

Gomory, Ralph E., Mathematical programming, Amer. Math. Monthly, 72 (1965) Suppl. pp. 99-110. [6.3.1]; p. 167.

────── and Hu, T.C., Multi-terminal flows in a network, SIAM J. Appl. Math., 9 (1961) 551-570; also in D.R. Fulkerson, Studies in Graph Theory, Part I, M.A.A., 1975, pp. 172-199. [6.3.2]; p. 168.

Good, I.J., Analogues of Poisson's summation formula, Amer. Math. Monthly, 69 (1962) 259-266. [6.4.3]; p. 175.

Good, R.A., Systems of linear relations, SIAM Review, 1 (1959) 1-31. [6.3.2]; p. 168.

Goodstein, R.L., Empiricism in mathematics, Dialectica, 23 (1969) 50-57. [2.2.0]; p. 43.

──────, Existence in mathematics, Compositio Math., 20 (1968) 70-82. [2.2.1]; p. 44.

Gordon, Richard, Herman, Gabor T. and Johnson, Steven A., Image reconstruction from projections, Scientific American, 233 (October 1975) 56-68, 139. [7.2.0]; p. 196.

Grabiner, Judith V., Is mathematical truth time-dependent?, Amer. Math. Monthly, 81 (1974) 354-365. [1.5.1]; p. 29.

Grattan-Guinness, Ivor, A mathematical union: William Henry and Grace Chesholm Young, Annals of Science, 29 (1972) 105-186. [1.3.0]; p. 17.

──────, Joseph Fourier, 1768-1830, M.I.T. Pr., 1972. [1.3.0]; p. 17.

──────, The Development of the Foundations of Mathematical Analysis from Euler to Riemann, M.I.T. Pr., 1970. [1.2.2]; p. 11.

──────, Towards a biography of Georg Cantor, Annals of Science, 27 (1971) 345-391. [1.3.0]; p. 17.

Graves, Lawrence M., Integration, McGraw-Hill Encyclopedia of Science and Technology, 1960, V. 7, pp. 166-174. [4.1.1]; p. 91.

──────, Nonlinear mappings between Banach spaces, in I.I. Hirschman, Jr., Studies in Real and Complex Analysis, M.A.A., 1965, pp. 34-54. [4.6.3]; p. 118.

──────, What is a functional?, Amer. Math. Monthly, 55 (1948) 467-472. [4.6.2]; p. 116.

Green, J.W., Recent applications of convex functions, Amer. Math. Monthly, 61 (1954) 449-454. [4.3.2]; p. 100.

Greenblatt, M.H., The "legal" value of π, and some related mathematical anomalies, Amer. Scientist, 53 (1965) 427A-434A. [1.2.0]; p. 4.

Greene, Francis A., *Combinations and permutations*, Encyclopedia Americana, 1976, V. 7, pp. 358-360. [3.1.0]; p. 60.

———, *Complex numbers*, Encyclopedia Americana, 1976, V. 7, pp. 460-461. [4.4.1]; p. 104.

Greenspan, Donald, *Computer power and its impact on applied mathematics*, in A.A. Taub, Studies in Applied Mathematics, M.A.A., 1971, pp. 65-89. [6.5.2]; p. 180.

Greenspan, H.P., *Applied mathematics as a science*, Amer. Math. Monthly, 68 (1961) 872-880. [1.4.1]; p. 23.

Greitzer, S.L. and Coxeter, H.S. MacDonald, Geometry Revisited, New Math. Libr., No. 19, Random House, 1967; M.A.A., 1975. [5.1.1]; p. 126.

Grenander, Ulf, *Computational probability and statistics*, SIAM Review, 15 (1973) 134-192. [6.1.1]; p. 156.

Gridgeman, N.T., *Geometric probability and the number pi*, Scripta Math., 25 (1960) 183-195. [6.1.2]; p. 157.

Griego, Richard J. and Hersh, Reuben, *Brownian motion and potential theory*, Scientific American, 220 (March 1969) 66-74, 148. [6.1.1]; p. 156.

Griffin, J.S., Jr., Boudreau, P.E. and Kac, Mark, *An elementary queueing problem*, Amer. Math. Monthly, 69 (1962) 713-724. [6.1.2]; p. 157.

Griffith, J.S., *Differential equations*, in N.J. Hardiman, Exploring University Mathematics, V. 2, Pergamon, 1968, pp. 99-115. [4.5.1]; p. 109.

Griffiths, Phillip A., *Periods of integrals on algebraic manifolds*, Bull. Amer. Math. Soc., 76 (1970) 228-296. [5.4.3]; p. 137.

Grinstein, Louise, *Determinant*, Encyclopedia Americana, 1976, V. 9, pp. 22-23. [3.4.1]; p. 75.

Grossman, Israel and Magnus, Wilhelm, Groups and Their Graphs, New Math. Libr., No. 14, Random House, 1964; M.A.A., 1975. [3.5.0]; p. 79.

Grünbaum, Adolf, *Geometry, chronometry and empiricism*, in H. Feigel and G. Maxwell, Minnesota Studies in the Philosophy of Science, V. 3, U. Minn. Pr., 1962, pp. 405-526. [2.2.1]; p. 45.

———, *Reply to Hilary Putnam's 'An examination of Grünbaum's philosophy of geometry'*, in R.S. Cohen and M.W. Wartofsky, Boston Studies in the Philosophy of Science, V. 5, D. Reidel, 1969, pp. 1-150. [2.2.1]; p. 45.

Grünbaum, Branko, *Polygons in arrangements generated by n points*, Math. Magazine, 46 (1973) 113-119. [3.1.1]; p. 60.

———, *Polytopal graphs*, in D.R. Fulkerson, Studies in Graph Theory, Part II, M.A.A., 1975, pp. 201-224. [3.2.3]; p. 65.

—— and Bose, Raj C., *Combinatorics and combinatorial geometry*, Encyclopaedia Britannica, 15th ed., 1974, Macropaedia, V. 4, pp. 942-954. [5.2.2]; p. 131.

—— and Shephard, G.C., *Convex polytopes*, Bull. London Math. Soc., 1 (1969) 257-300. [5.2.3]; p. 132.

Gugenheim, V.K.A.M., *Semisimplicial homotopy theory*, in P.J. Hilton, Studies in Modern Topology, M.A.A., 1968, pp. 99-133. [5.7.3]; p. 150.

Gulliksen, Harold, *Mathematical solutions for psychological problems*, Amer. Scientist, 47 (1959) 178-201. [7.3.1]; p. 202.

Haber, Seymour, *Numerical evaluation of multiple integrals*, SIAM Review, 12 (1970) 481-526. [6.4.2]; p. 174.

Hadamard, Jacques, The Psychology of Invention in the Mathematical Field, Princeton U. Pr., 1949. [1.5.0]; p. 26.

Haggett, Peter, *Network models in geography*, in R.J. Chorley and P. Haggett, Models in Geography, Methuen, 1967, pp. 609-668. [7.6.2]; p. 211.

Hahn, Hans, *Geometry and intuition*, Scientific American, 190 (April 1954) 84-91, 108; also in M. Kline, Mathematics in the Modern World, Freeman, 1968, pp. 184-188, 399. [5.1.0]; p. 125.

——, *Infinity*, in J.R. Newman, The World of Mathematics, V. 3, Simon and Schuster, 1956, pp. 1593-1611. [2.1.0]; p. 38.

——, *Is there an infinity?*, Scientific American, 187 (November 1952) 76-84, 104. [2.1.0]; p. 38.

——, *The crisis in intuition*, in J.R. Newman, The World of Mathematics, V. 3, Simon and Schuster, 1956, pp. 1956-1976. [1.5.0]; p. 26.

Hale, Jack K. and LaSalle, Joseph P., *Differential equations: linearity vs. nonlinearity*, SIAM Review, 5 (1963) 249-272. [4.5.2]; p. 109.

Hall, G.G., *An introduction to information theory*, Math. Teaching, 40 (1976) 4-7. [7.5.0]; p. 208.

Hall, Marshall, Jr., *Generators and relations in groups--the Burnside problem*, in T.L. Saaty, Lectures on Modern Mathematics, V. 2, Wiley, 1964, pp. 42-92. [3.5.2]; p. 80.

—— and Knuth, Donald E., *Combinatorial analysis and computers*, Amer. Math. Monthly, 72 (1965) Suppl. 21-28. [3.1.1]; p. 60.

Hall, P., *Some word problems*, J. London Math. Soc., 33 (1958) 482-496. [3.5.2]; p. 80.

Halmos, Paul R., *A glimpse into Hilbert space*, in T.L. Saaty, Lectures on Modern Mathematics, V. 1, Wiley, 1963, pp. 1-22. [4.6.2]; p. 116.

——, *Finite-dimensional Hilbert spaces*, Amer. Math. Monthly, 77 (1970) 457-464. [3.4.2]; p. 76.

——, *How to talk mathematics*, Notices Amer. Math. Soc., 21 (1974) 155-158. [1.6.0]; p. 32.

——, *How to write mathematics*, L'Enseignement Math., 16 (1970) 123-152. [1.6.0]; p. 32.

——, *Mathematics as a creative art*, Amer. Scientist, 56 (1968) 375-389. [1.5.0]; p. 26.

——, *"Nicolas Bourbaki"*, Scientific American, 196 (May 1957) 88-99, 174. [1.3.0]; p. 17.

——, *Recent progress in ergodic theory*, Bull. Amer. Math. Soc., 67 (1961) 70-80. [4.7.3]; p. 123.

——, *The basic concepts of algebraic logic*, Amer. Math. Monthly, 63 (1956) 363-387. [2.6.3]; p. 58.

——, *The foundations of probability*, Amer. Math. Monthly, 51 (1944) 493-510. [6.1.2]; p. 157.

——, *The legend of John von Neumann*, Amer. Math. Monthly, 80 (1973) 382-394. [1.3.0]; p. 17.

——, *What does the spectral theorem say?*, Amer. Math. Monthly, 70 (1963) 241-247. [4.6.3]; p. 118.

Hammersley, J.M., *Some speculations of a sense of nicely calculated chances*, SIAM Review, 16 (1974) 237-255. [6.1.1]; p. 156.

——, *Stochastic models for the distribution of particles in space*, Adv. Appl. Prob., 4 (1972) Suppl. pp. 44-68. [7.1.2]; p. 188.

Hamming, Richard W., *Impact of computers*, Amer. Math. Monthly, 72 (1965) Suppl. pp. 1-7. [6.5.0]; p. 177.

——, *Intellectual implications of the computer revolution*, Amer. Math. Monthly, 70 (1963) 4-11; also in T.L. Saaty and F.J. Weyl, The Spirit and Uses of the Mathematical Sciences, McGraw-Hill, 1969, pp. 188-199; in Studies in Mathematics, V. 16, SMSG, 1967, pp. 45-52; and in Z.W. Pylyshyn, Perspectives on the Computer Revolution, Prentice-Hall, 1970, pp. 370-377. [6.5.0]; p. 177.

——, *One man's view of computer science*, J. Assoc. Comp. Mach., 16 (1969) 3-12. [6.5.0]; p. 177.

Hammond, Allen L., *Sporadic groups: exceptions, or part of a pattern?*, Science, 181 (1973) 146, 148. [3.5.0]; p. 79.

Hanson, Norwood Russell, *Number theory and physical theory: an analogy*, in R.S. Cohen and M.W. Wartofsky, Boston Studies in the Philosophy of Science, V. 2, Humanities Pr., 1965, pp. 93-119. [7.1.1]; p. 186.

Harary, Frank, *Graph theory*, McGraw-Hill Encyclopedia of Science and Technology, 1960, V. 6, pp. 253-256. [3.2.1]; p. 63.

——, *On the history of the theory of graphs*, in F. Harary, New Directions in the Theory of Graphs, Academic Pr., 1973, pp. 1-17. [3.2.1]; p. 63.

——, *Some historical and intuitive aspects of graph theory*, <u>SIAM Review</u>, 2 (1960) 123-131. [3.2.0]; p. 63.

—— and Moser, Leo, *The theory of round robin tournaments*, <u>Amer. Math. Monthly</u>, 73 (1966) 231-246. [3.1.2]; p. 61.

Hardy, G.H., <u>A Mathematician's Apology</u>, Cambridge U. Pr., 1940; 1967; excerpted in J.R. Newman, <u>The World of Mathematics</u>, V. 4, Simon and Schuster, 1956, pp. 2027-2038. [1.5.0]; p. 26.

——, *An introduction to the theory of numbers*, <u>Bull. Amer. Math. Soc.</u>, 35 (1929) 778-818. [3.3.2]; p. 70.

——, *The integral $\int_0^\infty (\sin x/x)dx$*, <u>Math. Gazette</u>, 5 (1909); reprinted, ibid., 55 (1971) 152-158. [4.2.2]; p. 97.

Harlow, F.H., *Numerical fluid dynamics*, <u>Amer. Math. Monthly</u>, 72 (1965) Suppl. pp. 84-91. [7.1.1]; p. 186.

—— and Fromm, J.E., *Computer experiments in fluid dynamics*, <u>Scientific American</u>, 212 (March 1965) 104-110, 140. [7.1.0]; p. 184.

Harper, L.H. and Rota, Gian Carlo, *Matching theory: an introduction*, <u>Adv. in Prob.</u>, 1 (1971) 169-215. [3.1.2]; p. 61.

Harris, Zellig, *Mathematical linguistics*, in <u>The Mathematical Sciences--A Collection of Essays for COSRIMS</u>, M.I.T., 1969, pp. 190-196. [7.6.0]; p. 210.

Hashisaki, Joseph and Stoll, Robert R., *Set theory*, <u>Encyclopaedia Britannica</u>, 15th ed., 1974, Macropaedia V. 16, pp. 569-575. [2.1.1]; p. 39.

Hastings, S.P., *Some mathematical problems from neurobiology*, <u>Amer. Math. Monthly</u>, 82 (1975) 881-895. [7.2.2]; p. 198.

Hausner, Melvin, *Equation*, <u>Encyclopedia Americana</u>, 1976, V. 10, pp. 527-530. [3.6.1]; p. 85.

Hawkins, Thomas, *Cauchy and the spectral theory of matrices*, <u>Historia Math.</u>, 2 (1975) 1-29. [1.2.2]; p. 11.

——, *Hypercomplex numbers, Lie groups and the creation of group representation theory*, <u>Arch. Hist. Exact Sci.</u>, 8 (1972) 243-287. [1.2.3]; p. 13.

——, *Lebesgue's Theory of Integration: Its Origins and Development*, U. of Wisc. Pr., 1970; Chelsea, 1975. [1.2.2]; p. 11.

——, *New light on Frobenius' creation of the theory of group characters*, <u>Arch. Hist. Exact Sci.</u>, 12 (1974) 217-243. [1.2.3]; p. 13.

——, *The origins of the theory of group characters*, <u>Arch. Hist. Exact Sci.</u>, 7 (1971) 142-170. [1.2.3]; p. 13.

——, *The theory of matrices in the 19th century*, in <u>Proc. Inter. Cong. Math.</u> (1974), V. 2, Canad. Cong. Math., 1975, pp. 561-570. [1.2.2]; p. 11.

Heath, F.G., *Origins of the binary code*, Scientific American, 227 (August 1972) 76-83, 124. [6.5.0]; p. 177.

Helitzer, Florence, *A conversation with three mathematicians*, University: A Princeton Quarterly, 59 (Winter 1974) 1-5, 28-30. [1.5.0]; p. 26.

Henkin, Leon, *Are logic and mathematics identical?*, Science, 138 (1962) 788-794. [2.6.0]; p. 55.

———, *Mathematical foundations for mathematics*, Amer. Math. Monthly, 78 (1971) 463-487. [2.2.2]; p. 45.

———, *On mathematical induction*, Amer. Math. Monthly, 67 (1960) 323-338. [2.6.1]; p. 55.

Henrici, Peter, *Reflections of a teacher of applied mathematics*, Quarterly of Applied Math., 30 (1972) 31-39. [1.6.1]; p. 34.

Herman, Gabor T., Gordon, Richard and Johnson, Steven A., *Image reconstruction from projections*, Scientific American, 233 (October 1975) 56-68, 139. [7.2.0]; p. 196.

——— and Vitány, Paul M.B., *Growth functions associated with biological development*, Amer. Math. Monthly, 83 (1976) 1-15. [7.2.2]; p. 198.

Hermann, Robert, *E. Cartan's geometric theory of partial differential equations*, Advances in Math., 1 (1965) 265-317. [4.5.3]; p. 112.

Hermes, Henry, *A survey of recent results in differential equations*, SIAM Review, 15 (1973) 453-468. [4.5.3]; p. 112.

Hersh, Reuben, *How to classify differential polynomials*, Amer. Math. Monthly, 80 (1973) 641-654. [4.5.3]; p. 112.

——— and Cohen, Paul J., *Non-Cantorian set theory*, Scientific American, 217 (December 1967) 104-116, 160; also in M. Kline, Mathematics in the Modern World, Freeman, 1968, pp. 212-220, 400. [2.1.1]; p. 39.

——— and Davis, Martin, *Hilbert's 10th problem*, Scientific American, 229 (November 1973) 84-91, 136. [3.3.1]; p. 68.

——— and ———, *Nonstandard analysis*, Scientific American, 226 (June 1972) 78-86, 136. [2.4.1]; p. 49.

——— and Griego, Richard J., *Brownian motion and potential theory*, Scientific American, 220 (March 1969) 66-74, 148. [6.1.1]; p. 156.

Herz, Carl S., *Fourier series and integrals*, McGraw-Hill Encyclopedia of Science and Technology, 1960, V. 5, pp. 487-490. [4.2.1]; p. 95.

Hestenes, Magnus R., *An elementary introduction to the calculus of variations*, Math. Magazine, 23 (1950) 249-267. [4.7.1]; p. 121.

———, *Elements of the calculus of variations*, in E.F. Beckenbach, Modern Mathematics for the Engineer, McGraw-Hill, 1956, pp. 59-91. [4.7.2]; p. 121.

Hewitt, Edwin, *The role of compactness in analysis*, Amer. Math. Monthly, 67 (1960) 499-516. [4.3.2]; p. 100.

Heyting, Arend, *Disputation*, in P. Benacerraf and H. Putnam, Philosophy of Mathematics, Prentice-Hall, 1964, pp. 55-65. [2.3.0]; p. 47.

——, *Intuitionistic views on the nature of mathematics*, Synthese, 27 (1974) 79-91. [2.3.1]; p. 47.

——, *The intuitionist foundations of mathematics*, in P. Benacerraf and H. Putnam, Philosophy of Mathematics, Prentice-Hall, 1964, pp. 42-49. [2.3.0]; p. 47.

Hilbert, David, *On the infinite*, in P. Benacerraf and H. Putnam, Philosophy of Mathematics, Prentice-Hall, 1964, pp. 134-151. [2.1.1]; p. 39.

—— and Cohn-Vossen, Stephan, Geometry and the Imagination, Chelsea, 1952. [5.5.1]; p. 140.

Hildebrandt, T.H., *Integration in abstract spaces*, Bull. Amer. Math. Soc., 59 (1953) 111-139. [4.3.3]; p. 102.

——, *The Borel theorem and its generalizations*, Bull. Amer. Math. Soc., 32 (1926) 423-474. [4.3.3]; p. 102.

Hille, Einar, *Gelfond's solution of Hilbert's seventh problem*, Amer. Math. Monthly, 49 (1942) 654-661. [3.3.2]; p. 70.

——, *Topics in classical analysis*, in T.L. Saaty, Lectures on Modern Mathematics, V. 3, Wiley, 1965, pp. 1-57. [4.6.3]; p. 118.

——, *What is a semi-group?*, in I.I. Hirschman, Jr., Studies in Real and Complex Analysis, M.A.A., 1965, pp. 55-66. [4.6.3]; p. 118.

Hilton, Peter J., *Categories and functors*, Math. Teaching, 50 (1970) 30-34. [3.6.2]; p. 86.

——, *Introduction: modern topology*, in P.J. Hilton, Studies in Modern Topology, M.A.A., 1968, pp. 1-22. [5.7.3]; p. 150.

——, *Localization in topology*, Amer. Math. Monthly, 82 (1975) 113-131. [5.7.3]; p. 150.

——, *The art of mathematics*, Univ. of Birmingham, 1960. [1.5.0]; p. 27.

——, *The survival of education*, Educ. Tech., 13:11 (November 1973) 12-16. [1.6.0]; p. 32.

——, *Topology in the high school*, Educ. Studies Math., 3 (1970-71) 436-453. [5.7.2]; p. 149.

Hirolett, J., *Charles Babbage and his computer*, Math. Spectrum, 7 (1974-75) 73-80. [1.3.0]; p. 17.

Hirschman, I.I., Jr. and Widder, D.V., *The Laplace transform, the Stieltjes transform, and their generalizations*, in I.I. Hirschman, Jr., Studies in Real and Complex Analysis, M.A.A., 1965, pp. 67-89. [4.2.2]; p. 97.

Hoaglin, David C. and Mosteller, Frederick, *Statistics*, Encyclopaedia Britannica, 15th ed., 1974, Macropaedia V. 17, pp. 615-624. [6.2.1]; p. 162.

Hoffer, William, *A magic ratio recurs throughout art and nature*, Smithsonian, 6 (December 1975) 110-124. [3.3.0]; p. 68.

Hoffman, Alan J., *Eigenvalues of graphs*, in D.R. Fulkerson, Studies in Graph Theory, Part II, M.A.A., 1975, pp. 225-245. [3.2.3]; p. 65.

———, *Linear programming*, McGraw-Hill Encyclopedia of Science and Technology, 1960, V. 7, pp. 522-523. [6.3.1]; p. 167.

Hoffman, Banesh, Albert Einstein, Creator and Rebel, Viking, 1972. [1.3.0]; p. 17.

Hoffman, William C., *Visual illusions of angle as an application of Lie transformation groups*, SIAM Review, 13 (1971) 169-184. [7.2.2]; p. 198.

Holden, Alan, Shapes, Space and Symmetry, Columbia U. Pr., 1971. [5.1.0]; p. 125.

Honsberger, Ross, Ingenuity in Mathematics, New Math. Libr., No. 23, Random House, 1970; M.A.A., 1975. [3.1.1]; p. 60.

———, Mathematical Gems, M.A.A., 1973. [3.1.1]; p. 60.

Horváth, John, *An introduction to distributions*, Amer. Math. Monthly, 77 (1970) 227-240. [4.6.2]; p. 116.

Hotelling, Harold, *The statistical method and the philosophy of science*, Amer. Statistician, 12:5 (1958) 9-14. [6.2.0]; p. 161.

Householder, Alston S., *Generation of errors in digital computation*, Bull. Amer. Math. Soc., 60 (1954) 234-247. [6.4.1]; p. 173.

———, *Numerical analysis*, in T.L. Saaty, Lectures on Modern Mathematics, V. 1, Wiley, 1963, pp. 59-97. [6.4.2]; p. 174.

Hu, T.C. and Gomory, Ralph E., *Multi-terminal flows in a network*, SIAM J. Appl. Math., 9 (1961) 551-570; also in D.R. Fulkerson, Studies in Graph Theory, Part I, M.A.A., 1975, pp. 172-199. [6.3.2]; p. 168.

Huff, Darrell, How to Take a Chance, W.W. Norton, 1959. [6.1.0]; p. 154.

Huntley, H.E., The Divine Proportion: A Study in Mathematical Beauty, Dover, 1970. [1.7.0]; p. 36.

Hyers, D.H., *Linear topological spaces*, Bull. Amer. Math. Soc., 51 (1945) 1-21. [4.6.2]; p. 116.

Iliev, L., *Mathematics as the science of models*, Russian Math. Surveys, 27:2 (1972) 181-189. [1.5.0]; p. 27.

Infeld, Leopold, Whom the Gods Love, Whittlesey House, 1948. [1.3.0]; p. 17.

Isard, Walter and Kaniss, Phyllis, *The 1973 Nobel prize for economic science*, Science, 182 (1973) 568-569, 571. [7.4.0]; p. 204.

Iwasawa, Kenkichi and Narasimhan, Raghavan, *Complex analysis*, <u>Encyclopaedia Britannica</u>, 15th ed., 1974, Macropaedia V. 1, pp. 719-735. [4.4.2]; p. 105.

Iyanaga, Shokichi, *Algebraic theory of numbers*, in A.H. Livermore, <u>Science in Japan</u>, AAAS, 1965, pp. 81-113. [3.3.3]; p. 72.

Jackson, Dunham, *Orthogonal trigonometric sums*, <u>Annals of Math.</u>, 34 (1933) 799-814. [4.3.3]; p. 102.

———, *Series of orthogonal polynomials*, <u>Annals of Math.</u>, 34 (1933) 527-545. [4.3.3]; p. 102.

———, *The convergence of Fourier series*, <u>Amer. Math. Monthly</u>, 41 (1934) 67-84. [4.3.2]; p. 100.

James, R.D., *Recent progress in the Goldbach problem*, <u>Bull. Amer. Math. Soc.</u>, 55 (1949) 246-260. [3.3.2]; p. 70.

Jeffreys, Harold, *The present position in probability theory*, <u>Brit. J. Phil. Sci.</u>, 5 (1955) 275-289. [6.1.2]; p. 158.

Johnson, Norman L. and Kotz, Samuel, *Statistical distributions: a survey of the literature, trends and prospects*, <u>Amer. Statistician</u>, 27 (1973) 15-17. [6.2.2]; p. 162.

Johnson, Paul and Redheffer, Raymond, *Scrambled series*, <u>Amer. Math. Monthly</u>, 73 (1966) 822-828. [4.2.2]; p. 97.

Johnson, Steven A., Gordon, Richard and Herman, Gabor T., *Image reconstruction from projections*, <u>Scientific American</u>, 233 (October 1975) 56-68, 139. [7.2.0]; p. 196.

Jones, Burton W., *Theory of numbers*, <u>Encyclopedia Americana</u>, 1976, V. 20, pp. 538-541. [3.3.1]; p. 68.

Jones, D.S., *Asymptotic behavior of integrals*, <u>SIAM Review</u>, 14 (1972) 286-317. [4.7.2]; p. 121.

Jones, James P., *Recursive undecidability--an exposition*, <u>Amer. Math. Monthly</u>, 81 (1974) 724-738. [2.5.2]; p. 53.

Jones, Landon Y., Jr., *Bad days on Mount Olympus*, <u>Atlantic Monthly</u>, 233 (February 1974) 37-46, 51-53. [1.7.0]; p. 36.

———, *Mathematicians: They're special*, <u>Think</u>, 40:4 (1974) 32-35. [1.5.0]; p. 27.

Jones, Phillip S., *The history of mathematical education*, <u>Amer. Math. Monthly</u>, 74 (1967) Suppl. pp. 38-55. [1.6.0]; p. 32.

Joyce, D.C., *Survey of extrapolation processes in numerical analysis*, <u>SIAM Review</u>, 13 (1971) 435-490. [6.4.2]; p. 174.

Kac, Mark, *Can one hear the shape of a drum?*, <u>Amer. Math. Monthly</u>, 73 (1966) Suppl. pp. 1-23. [7.1.2]; p. 188.

———, *Hugo Steinhaus--a reminiscence and a tribute*, <u>Amer. Math. Monthly</u>, 81 (1974) 572-581. [1.3.0]; p. 17.

———, *On applying mathematics: reflections and examples*, Quarterly of Applied Math., 30 (1972) 17-29. [1.5.3]; p. 30.

———, *Probability*, Scientific American, 211 (September 1964) 92-108, 269; also in M. Kline, Mathematics in the Modern World, Freeman, 1968, pp. 165-174, 398-399; in D. Messick, Mathematical Thinking in Behavioral Sciences, Freeman, 1968, pp. 23-32, 223-224; and in The Mathematical Sciences--A Collection of Essays for COSRIMS, M.I.T., 1969, pp. 232-251. [6.1.0]; p. 154.

———, *Random walk and the theory of Brownian motion*, Amer. Math. Monthly, 54 (1947) 369-391. [6.1.2]; p. 158.

———, *Some mathematical models in science*, Science, 166 (1969) 695-699. [7.2.2]; p. 198.

———, *Will computers replace humans?*, in E.H. Kone and H.J. Jordan, The Greatest Adventure, Rockefeller U. Pr., 1974, pp. 193-206. [6.5.0]; p. 177.

———, Boudreau, P.E. and Griffin J.S., Jr., *An elementary queueing problem*, Amer. Math. Monthly, 69 (1962) 713-724. [6.1.2]; p. 157.

——— and Ulam, Stanislaw M., Mathematics and Logic: Retrospect and Prospects, Frederick A. Praeger, 1968; The New American Library, 1969. [1.1.2]; p. 3.

Kaniss, Phyllis and Isard, Walter, *The 1973 Nobel prize for economic science*, Science, 182 (1973) 568-569, 571. [7.4.0]; p. 204.

Kaplansky, Irving, *Lie algebras*, in T.L. Saaty, Lectures on Modern Mathematics, V. 1, Wiley, 1963, pp. 115-132. [3.4.3]; p. 76.

———, *Topological rings*, Bull. Amer. Math. Soc., 54 (1948) 809-826. [5.8.2]; p. 152.

———, et al., *Number theory*, Encyclopaedia Britannica, 15th ed., 1974, Macropaedia V. 13, pp. 358-381. [3.3.3]; p. 72.

Karlin, Samuel, *Some mathematical models of population genetics*, Amer. Math. Monthly, 79 (1972) 699-739. [7.2.2]; p. 198.

———, *The mathematical theory of inventory processes*, in E.F. Beckenbach, Modern Mathematics for the Engineer, 2nd Ser., McGraw-Hill, 1961, pp. 228-258. [6.3.2]; p. 168.

Kasner, Edward and Newman, James R., *Paradox lost and paradox regained*, in J.R. Newman, The World of Mathematics, V. 3, Simon and Schuster, 1956, pp. 1936-1955. [2.1.0]; p. 38.

Kato, Tosio, *Scattering theory*, in A.H. Taub, Studies in Applied Mathematics, M.A.A., 1971, pp. 90-115. [7.1.3]; p. 190.

Kaufman, Hyman, *Real analysis*, Encyclopaedia Britannica, 15th ed., 1974, Macropaedia V. 1, pp. 772-791. [4.3.1]; p. 99.

Kautz, William H. and Turner, James, *A survey of progress in graph theory in the Soviet Union*, SIAM Review, 12 (1970) Suppl. pp. 1-68. [3.2.2]; p. 64.

Kay, David C., *Non-Euclidean geometry*, Encyclopedia Americana, 1976, V. 12, pp. 494-499. [5.1.1]; p. 127.

Kazarinoff, Nicholas D., Geometric Inequalities, New Math. Libr., No. 4, Random House, 1961; M.A.A., 1975. [6.3.0]; p. 165.

Keeping, E.S., *Statistical decisions*, Amer. Math. Monthly, 63 (1956) 147-159. [6.2.1]; p. 162.

Keisler, H. Jerome, *A survey of ultraproducts*, in Y. Bar-Hillel, Logic, Methodology and Philosophy of Science, North-Holland, 1965, pp. 112-126. [2.6.3]; p. 58.

——, *Forcing and the omitting types theorem*, in M.D. Morley, Studies in Model Theory, M.A.A., 1973, pp. 96-133. [2.6.3]; p. 58.

——, *Model theory*, in Actes Cong. Inter. Math. (1970), V. 1, Gauthier-Villars, 1971, pp. 141-150. [2.6.2]; p. 56.

—— and Tarski, Alfred, *From accessible to inaccessible cardinals*, Fund. Math., 53 (1963) 225-308. [2.1.3]; p. 40.

Keldyš, M.V., *Functions of a complex variable*, in A.D. Aleksandrov, et al., Mathematics--Its Content, Methods and Meaning, V. 2, M.I.T., 1963, pp. 139-195. [4.4.2]; p. 105.

Keller, Joseph B., *Inverse problems*, Amer. Math. Monthly, 83 (1976) 107-118. [7.1.2]; p. 188.

—— and McLaughlin, D.W., *The Feynman integral*, Amer. Math. Monthly, 82 (1975) 451-465. [4.7.3]; p. 123.

Kelly, Paul J., *Plane convex figures*, NCTM Twenty-Eighth Yearbook, 1963, 251-264. [5.2.0]; p. 131.

Kemeny, John G., *Mathematical models and the computer*, Pi Mu Epsilon Journal, 5 (1973) 373-386. [6.5.0]; p. 177.

——, *Teaching the new mathematics*, Atlantic Monthly, October 1962; also in J.G. Kemeny, Random Essays on Mathematics, Education and Computers, Prentice-Hall, 1964, pp. 27-34. [1.6.0]; p. 32.

——, *The social sciences call on mathematics*, in The Mathematical Sciences--A Collection of Essays for COSRIMS, M.I.T., 1969, pp. 21-36. [7.3.0]; p. 201.

——, *What every college president should know about mathematics*, Amer. Math. Monthly, 80 (1973) 889-901. [7.6.0]; p. 210.

Kempner, A.J., *Remarks on 'unsolvable' problems*, Amer. Math. Monthly, 43 (1936) 467-473. [2.5.1]; p. 53.

Kendall, David G., *Branching processes since 1873*, J. London Math. Soc., 41 (1966) 385-406. [6.1.2]; p. 158.

——, *The genealogy of genealogy: branching processes before (and after) 1873*, Bull. London Math. Soc., 7 (1975) 225-253. [6.1.2]; p. 158.

Kestelman, H., *Wallpaper patterns*, in N.J. Hardiman, Exploring University Mathematics, V. 2, Pergamon, 1968, pp. 60-85. [3.5.0]; p. 79.

Kiefer, Jack, *Statistical inference*, in The Mathematical Sciences--A Collection of Essays for COSRIMS, M.I.T., 1969, pp. 60-71; also in Math. Spectrum, 3 (1970-1971) 1-11. [6.2.0]; p. 161.

Kiernan, B. Melvin, *The development of Galois theory from Lagrange to Artin*, Arch. Hist. Exact Sci., 8 (1971) 40-154. [1.2.1]; p. 8.

Kimberling, Clark H., *Emmy Noether*, Amer. Math. Monthly, 79 (1972) 136-149; Addendum, 79 (1972) 755. [1.3.0]; p. 17.

King, Gilbert W., *Applied mathematics in operations research*, in E.F. Beckenbach, Modern Mathematics for the Engineer, McGraw-Hill, 1956, pp. 211-242. [6.3.1]; p. 167.

Kingman, J.F.C., *Markov population processes*, J. Appl. Prob., 6 (1969) 1-18. [7.2.3]; p. 199.

Klamkin, Murray S., *On the ideal role of an industrial mathematician and its educational implications*, Educ. Studies Math., 3 (1970-71) 244-269; also in Amer. Math. Monthly, 78 (1971) 53-76. [1.6.0]; p. 33.

———, *The teaching of mathematics so as to be useful*, Educ. Studies Math., 1 (1968-69) 126-160. [1.6.1]; p. 34.

——— and Newman, D.J., *The philosophy and applications of transform theory*, SIAM Review, 3 (1961) 10-36. [7.1.2]; p. 189.

Klee, Victor, *The Euler characteristic in combinatorial geometry*, Amer. Math. Monthly, 70 (1963) 119-127. [5.2.3]; p. 132.

———, *What is a convex set?*, Amer. Math. Monthly, 78 (1971) 616-631. [5.2.1]; p. 131.

Kleene, Stephen C., *The new logic*, Amer. Scientist, 57 (1969) 333-347. [2.5.1]; p. 53.

——— and Feferman, Solomon, *Foundations of mathematics*, Encyclopaedia Britannica, 15th ed., 1974, Macropaedia V. 11, pp. 630-639. [2.2.1]; p. 45.

Kleiman, S.L. and Laksov, Dan, *Schubert calculus*, Amer. Math. Monthly, 79 (1972) 1061-1082. [5.4.3]; p. 137.

Klein, Lawrence R., *The role of mathematics in economics*, in The Mathematical Sciences--A Collection of Essays for COSRIMS, M.I.T., 1969, pp. 161-175. [7.4.1]; p. 204.

Kline, J.R., *What is the Jordan curve theorem?*, Amer. Math. Monthly, 49 (1942) 281-286. [5.5.1]; p. 140.

Kline, Morris, *Calculus*, Encyclopedia Amerciana, 1976, V. 5, pp. 163-177. [4.1.1]; p. 92.

———, *Differential equations*, Encyclopedia Americana, 1976, V. 9, pp. 109-110. [4.5.1]; p. 109.

———, *Geometry: history and development*, Encyclopedia Americana, 1976, V. 12, pp. 471-478. [1.2.1]; p. 8.

———, *Logic versus pedagogy*, Amer. Math. Monthly, 77 (1970) 264-282. [1.6.1]; p. 34.

———, Mathematical Thought from Ancient to Modern Times, Oxford U. Pr., 1972. [1.2.1]; p. 8.

———, Mathematics in Western Culture, Oxford U. Pr., 1953. [1.1.0]; p. 1.

Knebelman, M.S., *Graphical coordinate systems*, McGraw-Hill Encyclopedia of Science and Technology, 1960, V. 3, pp. 454-456. [5.1.0]; p. 125.

———, et al., *Geometry*, McGraw-Hill Encyclopedia of Science and Technology, 1960, V. 6, pp. 150-164. [5.1.1]; p. 127.

Kneebone, G.T., *Logic*, in N.J. Hardiman, Exploring University Mathematics, V. 3, Pergamon, 1969, pp. 68-79. [2.6.0]; p. 55.

Knuth, Donald E., *Ancient Babylonian algorithms*, Comm. Assoc. Comp. Mach., 15 (1972) 671-677. [1.2.0]; p. 4.

———, *Computer programming as an art*, Comm. Assoc. Comp. Mach., 17 (1974) 667-673. [6.5.0]; p. 178.

———, *Computer science and its relation to mathematics*, Amer. Math. Monthly, 81 (1974) 323-343. [6.5.1]; p. 179.

———, *Computer science and mathematics*, Amer. Scientist, 61 (1973) 707-713. [6.5.1]; p. 179.

———, *George Forsythe and the development of computer science*, Comm. Assoc. Comp. Mach., 15 (1972) 721-726. [1.3.0]; p. 18.

——— and Hall, Marshall, Jr., *Combinatorial analysis and computers*, Amer. Math. Monthly, 72 (1965) Suppl. pp. 21-28. [3.1.1]; p. 60.

Koehler, J.E., *Folding a strip of stamps*, J. Combinatorial Theory, 5 (1968) 135-152. [3.1.2]; p. 61.

Kolata, Gina Bari, *Analysis of algorithms: coping with hard problems*, Science, 186 (1974) 520-521. [6.5.1]; p. 179.

———, *Cascading bifurcations: the mathematics of chaos*, Science, 189 (1975) 984-985. [7.1.1]; p. 186.

———, *Combinatorics: steps toward a unified theory*, Science, 183 (1974) 839-840, 883. [3.1.1]; p. 60.

———, *Foundations of mathematics: ties to infinite games*, Science, 188 (1975) 923-924. [2.1.1]; p. 39.

———, *Mathematical problems: a committee to replace Hilbert*, Science, 185 (1974) 430. [1.2.0]; p. 4.

———, *Riemann hypotheses: elusive zeros of the zeta functions*, Science, 185 (1974) 429-431. [4.4.1]; p. 104.

Kolman, Bernard and Belinfante, Johan G.F., *An introduction to Lie groups and Lie algebras, with applications*, I-III, SIAM Review, 8 (1966) 11-46; 10 (1968) 160-195; 11 (1969) 510-543; revised and reprinted as A Survey of Lie Groups and Lie Algebras, SIAM, 1972. [5.8.3]; p. 152.

Kolmogorov, A.N., *The theory of probability*, in A.D. Aleksandrov, et al., Mathematics--Its Content, Methods and Meaning, V. 2, M.I.T., 1963, pp. 229-264. [6.1.1]; p. 156.

Koopman, Bernard Osgood, *The axioms and algebra of intuitive probability*, Annals of Math., 41 (1940) 269-292. [6.1.2]; p. 158.

Koopmans, Tjalling C. and Bausch, Augustus F., *Selected topics in economics involving mathematical reasoning*, SIAM Review, 1 (1959) 79-148. [7.4.2]; p. 205.

Koppelman, Elaine, *The calculus of operations and the rise of abstract algebra*, Arch. Hist. Exact Sci., 8 (1971) 155-242. [1.2.1]; p. 8.

Koptsik, V.A. and Shubnikov, A.V., Symmetry in Science and Art, Plenum Pr., 1974. [3.5.1]; p. 79.

Köthe, G. and Ballier, F., *The changing structure of modern mathematics*, in H. Behnke, et al., Fundamentals of Mathematics, V. 3, M.I.T., 1974, pp. 505-528. [1.5.2]; p. 30.

Kotz, Samuel and Johnson, Norman L., *Statistical distributions: a survey of the literature, trends and prospects*, Amer. Statistician, 27 (1973) 15-17. [6.2.2]; p. 162.

Kramer, Edna E., The Nature and Growth of Modern Mathematics, Hawthorn, 1970; Fawcett, 1973. [1.1.0]; p. 1.

Krantz, David, *A survey of measurement theory*, in G.B. Dantzig and A.F. Veinott, Jr., Mathematics of the Decision Sciences, Part 2, Amer. Math. Soc., 1968, pp. 314-350. [7.3.2]; p. 202.

Kreisel, Georg, *Mathematical logic*, in T.L. Saaty, Lectures on Modern Mathematics, V. 3, Wiley, 1965, pp. 95-195. [2.6.2]; p. 56.

Kruskal, William, *Statistics, Molière, and Henry Adams*, Amer. Scientist, 55 (1967) 416-420. [6.2.0]; p. 161.

Krylov, V.I., *The calculus of variations*, in A.D. Aleksandrov, et al., Mathematics--Its Content, Methods and Meaning, V. 2, M.I.T., 1963, pp. 119-138. [4.7.1]; p. 121.

Kuhn, Harold W., *"Steiner's" problem revisited*, in G.B. Dantzig and B.C. Eaves, Studies in Optimization, M.A.A., 1974, pp. 52-70. [6.3.2]; p. 168.

────── and Tucker, Albert W., *John von Neumann's work in the theory of games and mathematical economics*, Bull. Amer. Math. Soc., 64 (1958) Suppl. pp. 100-122. [1.3.1]; p. 19.

Kuller, Robert G., *Coin tossing, probability and the Weierstrass approximation theorem*, Math. Magazine, 37 (1964) 262-265. [6.1.2]; p. 158.

Kuratowski, Kazimierz, *Wacław Sierpiński (1882-1969)*, Acta Arith., 21 (1972) 1-5. [1.3.0]; p. 18.

Kurosh, A.G., Livshits, A. Kh. and Shul'Geifer, E.G., *Foundations of the theory of categories*, Russian Math. Surveys, 15:6 (1960) 1-46. [3.6.2]; p. 86.

Kürschák, József, Hungarian Problem Book I, New Math. Libr., No. 11, Random House, 1963; M.A.A., 1975. [1.7.1]; p. 37.

──────, Hungarian Problem Book II, New Math. Libr., No. 12, Random House, 1963; M.A.A., 1975. [1.7.1]; p. 37.

Ladyzenskaja, O.A. and Sobolev, S.L., *Partial differential equations*, in A.D. Aleksandrov, et al., Mathematics--Its Content, Methods and Meaning, V. 2, M.I.T., 1963, pp. 3-55. [4.5.2]; p. 111.

Lagerstrom, P.A. and Casten, R.G., *Basic concepts underlying singular perturbation techniques*, SIAM Review, 14 (1972) 63-120. [7.1.3]; p. 190.

Lakatos, Imre, *Proofs and refutations*, Brit. J. Phil. Science, 14 (1963-64) 1-25, 120-139, 221-245, 296-342. [1.5.2]; p. 30.

Laksov, Dan and Kleiman, S.L., *Schubert calculus*, Amer. Math. Monthly, 79 (1972) 1061-1082. [5.4.3]; p. 137.

Lam, T.Y. and Siu, M.K., K_0 and K_1--*an introduction to algebraic K-theory*, Amer. Math. Monthly, 82 (1975) 329-364. [3.6.3]; p. 88.

Lambek, Joachim, *The mathematics of sentence structures*, Amer. Math. Monthly, 65 (1958) 154-170. [7.6.2]; p. 212.

Lanczos, Cornelius, *Linear systems in self-adjoint form*, Amer. Math. Monthly, 65 (1958) 665-679. [4.5.2]; p. 110.

——, Space Through the Ages, Academic Pr., 1970. [5.1.1]; p. 127.

——, *William Rowan Hamilton--an appreciation*, Amer. Scientist, 55 (1967) 129-143. [1.3.0]; p. 18.

Lange, L.H. and Chakerian, G.D., *Geometric extremum problems*, Math. Magazine, 44 (1971) 57-69. [6.3.1]; p. 166.

Langer, R.E., *Fourier's series--the genesis and evolution of a theory*, Amer. Math. Monthly, 54 (1947) Suppl. pp. 1-86. [1.2.1]; p. 8.

——, *What are Eigen-werte?*, Amer. Math. Monthly, 50 (1943) 279-287. [3.4.1]; p. 75.

Langford, Eric, *A problem in geometric probability*, Math. Magazine, 43 (1970) 237-244. [6.1.2]; p. 158.

LaSalle, Joseph P. and Hale, Jack K., *Differential equations: linearity vs. nonlinearity*, SIAM Review, 5 (1963) 249-272. [4.5.2]; p. 109.

Lashof, Richard, *The tangent bundle of a topological manifold*, Amer. Math. Monthly, 79 (1972) 1090-1096. [5.6.3]; p. 146.

Lass, Harry, *Calculus of vectors*, McGraw-Hill Encyclopedia of Science and Technology, 1960, V. 2, pp. 410-414. [4.2.1]; p. 95.

Lavrent'ev, M.A. and Nikol'skiĭ, S.M., *Analysis*, in A.D. Aleksandrov, et al., Mathematics--Its Content, Methods and Meaning, V. 1, M.I.T., 1963, pp. 65-180. [4.1.1]; p. 92.

Lawson, H. Blaine, *Foliations*, Bull. Amer. Math. Soc., 80 (1974) 369-418. [5.6.3]; p. 147.

Lax, Peter D., *Numerical solution of partial differential equations*, Amer. Math. Monthly, 72 (1965) Suppl. pp. 74-84. [4.5.2]; p. 110.

——, *The formation and decay of shock waves*, Amer. Math. Monthly, 79 (1972) 227-241. [4.5.2]; p. 110.

Layzer, David, *The arrow of time*, Scientific American, 233 (December 1975) 56-69, 148. [7.1.1]; p. 186.

Lazarus, Mitchell, *Mathophobia: some personal speculations*, Nat. Elem. Principal, 53:2 (Jan.-Feb. 1974) 16-22. [1.6.0]; p. 33.

Lebesgue, Henri, *The development of the integral concept*, in H. Lebesgue, Measure and the Integral, Holden-Day, 1966, pp. 178-194. [4.3.2]; p. 100.

Le Corbeiller, P., *The curvature of space*, Scientific American, 191 (November 1954) 80-86, 124; also in M. Kline, Mathematics in the Modern World, Freeman, 1968, pp. 128-133, 397. [1.2.0]; p. 4.

Lederberg, Joshua, *Topology of molecules*, in The Mathematical Sciences--A Collection of Essays for COSRIMS, M.I.T., 1969, pp. 37-51. [7.1.0]; p. 185.

Lefschetz, Solomon, *A page of mathematical autobiography*, Bull. Amer. Math. Soc., 74 (1968) 854-879. [5.4.3]; p. 138.

———, *Linear and nonlinear oscillations*, in E.F. Beckenbach, Modern Mathematics for the Engineer, McGraw-Hill, 1956, pp. 7-29. [4.5.1]; p. 109.

———, *Reminiscences of a mathematical immigrant in the United States*, Amer. Math. Monthly, 77 (1970) 344-350. [1.3.0]; p. 18.

———, *The early development of algebraic geometry*, Amer. Math. Monthly, 76 (1969) 451-460. [1.2.3]; p. 13.

———, *The structure of mathematics*, Amer. Scientist, 38 (1950) 105-111. [1.5.0]; p. 27.

Lehmer, D.H., *Computer technology applied to the theory of numbers*, in W.J. LeVeque, Studies in Number Theory, M.A.A., 1969, pp. 117-151. [3.3.2]; p. 70.

———, *Mechanized mathematics*, Bull. Amer. Math. Soc., 72 (1966) 739-750. [1.5.1]; p. 29.

Lehner, Joseph, *The Picard theorems*, Amer. Math. Monthly, 76 (1969) 1005-1012. [4.4.2]; p. 106.

Leibowitz, Martin A., *Queues*, Scientific American, 219 (August 1968) 96-103, 124. [6.1.0]; p. 154.

Leonard, J.L. and Bruckner, Andrew M., *Derivatives*, Amer. Math. Monthly, 73 (1966) Suppl. pp. 24-56. [4.1.2]; p. 92.

Leontief, Wassily, *Mathematics in economics*, Bull. Amer. Math. Soc., 60 (1954) 215-233. [7.4.0]; p. 204.

Leslie, P.H., *On the use of matrices in certain population mathematics*, Biometrika, 33 (1945) 183-212. [7.2.2]; p. 198.

LeVeque, William J., *A brief survey of diophantine equations*, in W. J. LeVeque, Studies in Number Theory, M.A.A., 1969, pp. 4-24. [3.3.2]; p. 71.

———, MacDuffee, C.C. and Smith, David E., *Arithmetic*, Encyclopaedia Britannica, 15th ed., 1974, Macropaedia V. 1, pp. 1171-1178. [3.6.0]; p. 84.

———, et al., *History of mathematics*, Encyclopaedia Britannica, 15th ed., 1974, Macropaedia, V. 11, pp. 639-670. [1.2.0]; p. 5.

Levins, Richard, *The strategy of model building in population biology*, Amer. Scientist, 54 (1966) 421-431. [7.2.1]; p. 197.

Levinson, Norman, *A motivated account of an elementary proof of the prime number theorem*, Amer. Math. Monthly, 76 (1969) 225-245. [3.3.2]; p. 71.

———, *Coding theory: a counterexample to G.H. Hardy's conception of applied mathematics*, Amer. Math. Monthly, 77 (1970) 249-258. [7.5.2]; p. 209.

Levitz, Hilbert, *Non-standard analysis: an exposition*, L'Enseignement Math., 20 (1974) 9-32. [2.4.1]; p. 49.

Lewis, D.J., *Diophantine equations: p-adic methods*, in W.J. LeVeque, Studies in Number Theory, M.A.A., 1969, pp. 25-75. [3.3.3]; p. 72.

Li, Ching Chun, *Biometrics*, McGraw-Hill Encyclopedia of Science and Technology, 1960, V. 2, pp. 223-232. [7.2.1]; p. 197.

Lieber, Lillian R., Galois and the Theory of Groups: A Bright Star in Mathesis, Galois Institute, 1932; 1956. [3.5.1]; p. 79.

———, Human Values and Science, Art and Mathematics, Norton, 1961. [1.4.0]; p. 21.

———, Infinity, Holt, Rinehart and Winston, 1953, 1964. [2.1.0]; p. 38.

———, Mits, Wits, and Logic, Norton, 1947; 1954, 1960. [1.7.0]; p. 37.

———, Non-Euclidean Geometry or Three Moons in Mathesis, Science Pr., 1940. [5.1.0]; p. 125.

———, Take a Number, Ronald Pr., 1946. [1.7.0]; p. 37.

———, The Education of T.C. Mits, Norton, 1942; 1944. [1.7.0]; p. 37.

———, The Einstein Theory of Relativity, Holt, Rinehart and Winston, 1936; 1945. [7.1.1]; p. 186.

Lietzmann, W., Visual Topology, American Elsevier, 1965. [5.5.0]; p. 139.

Lifshits, V.N. and Sadovskii, L.E., *Algebraic models of computing machines*, Russian Math. Surveys, 27:3 (1972) 87-135. [6.5.3]; p. 181.

Lighthill, M.J., *The art of teaching the art of applying mathematics*, Math. Gazette, 55 (1971) 249-270. [1.6.0]; p. 33.

Lightstone, A.H., *Infinitesimals*, Amer. Math. Monthly, 79 (1972) 242-251. [2.4.1]; p. 49.

———, *Infinitesimals and integration*, Math. Magazine, 46 (1973) 20-30. [2.4.2]; p. 50.

Lin, C.C., *Dynamics of self-gravitating systems--structure of galaxies*, SIAM Review, 11 (1969) 127-151; also in A.H. Taub, Studies in Applied Mathematics, M.A.A., 1971, pp. 116-149. [7.1.3]; p. 191.

Linear and multilinear algebra, Encyclopaedia Britannica, 15th ed., 1974, Macropaedia V. 1, pp. 507-518. [3.4.2]; p. 76.

Livshits, A. Kh., Kurosh, A.G. and Shul'Geifer, E.G., *Foundations of the theory of categories*, Russian Math. Surveys, 15:6 (1960) 1-46. [3.6.2]; p. 86.

Loève, Michel, *On stochastic processes*, in T.L. Saaty, Lectures on Modern Mathematics, V. 3, Wiley, 1965, pp. 245-276. [6.1.3]; p. 159.

Lorch, Edgar R., *The spectral theorem*, in R.C. Buck, Studies in Modern Analysis, M.A.A., 1962, pp. 88-137. [4.6.2]; p. 116.

———, *The structure of normed Abelian rings*, Bull. Amer. Math. Soc., 50 (1944) 447-463. [4.6.3]; p. 118.

Lorentz, G.G., *Metric entropy, widths, and superpositions of functions*, Amer. Math. Monthly, 69 (1962) 469-485. [6.4.3]; p. 176.

Lorenzen, P., *Constructive mathematics as a philosophical problem*, Compositio Math., 20 (1968) 133-142. [2.3.1]; p. 47.

Lowan, Arnold N., *Numerical analysis*, McGraw-Hill Encyclopedia of Science and Technology, 1960, V. 9, pp. 227-229. [6.4.1]; p. 173.

Lucas, William F., *An overview of the mathematical theory of games*, Management Science, 18 (1971-72) P3-P19. [6.3.2]; p. 169.

Luce, R. Duncan, *The mathematics used in mathematical psychology*, Amer. Math. Monthly, 71 (1964) 364-378; also in Studies in Mathematics, V. 16, SMSG, 1967, pp. 181-195. [7.3.1]; p. 202.

Luchins, Abraham S. and Luchins, Edith H., *Logicism*, Scripta Math., 27 (1965) 223-243. [2.2.0]; p. 43.

Luchins, Edith H. and Luchins, Abraham S., *Logicism*, Scripta Math., 27 (1965) 223-243. [2.2.0]; p. 43.

Luxemburg, W.A.J., *A general theory of monads*, in W.A.J. Luxemburg, Applications of Model Theory to Algebra, Analysis, and Probability, Holt, Rinehart and Winston, 1969, pp. 18-86. [2.4.3]; p. 50.

———, *What is nonstandard analysis*, Amer. Math. Monthly, 80 (1973) Suppl. pp. 38-67. [2.4.2]; p. 50.

Lyon, Thoburn C., *Projection*, Encyclopedia Americana, 1976, V. 22, pp. 644-650b. [5.1.1]; p. 127.

Lyubich, Yu. I., *Basic concepts and theorems of the evolutionary genetics of free populations*, Russian Math. Surveys, 26:5 (1971) 55-123. [7.2.2]; p. 198.

Lyusternik, L.A., *The early years of the Moscow mathematical school*, Russian Math. Surveys, 22:1 (1967) 133-157; 22:2 (1967) 171-211; 22:4 (1967) 55-91; 25:4 (1970) 167-174. [1.2.2]; p. 11.

MacDuffee, C.C., Smith, David E. and LeVeque, William J., *Arithmetic*, Encyclopaedia Britannica, 15th ed., 1974, Macropaedia V. 1, pp. 1171-1178. [3.6.0]; p. 84.

Mackey, George W., *Ergodic theory and its significance for statistical mechanics and probability theory*, Advances in Math., 12 (1974) 178-286. [4.7.3]; p. 123.

———, *Functions on locally compact groups*, Bull. Amer. Math. Soc., 56 (1950) 385-412. [4.7.3]; p. 123.

———, *Group representations and analysis*, Rice Univ. Studies, 49:4 (1963) 13-27. [5.8.2]; p. 152.

———, *Group theory and its significance for mathematics and physics*, Proc. Amer. Phil. Soc., 117 (1973) 374-380. [3.5.0]; p. 79.

———, *Infinite-dimensional group representations*, Bull. Amer. Math. Soc., 69 (1963) 628-686. [5.8.3]; p. 152.

———, *Operator theory*, McGraw-Hill Encyclopedia of Science and Technology, 1960, V. 9, pp. 338-341. [4.6.2]; p. 116.

———, *Quantum mechanics and Hilbert space*, Amer. Math. Monthly, 64 (1957) Suppl. pp. 45-57. [7.1.2]; p. 189.

Mackie, A.G., *Some comments on existence and uniqueness theorems in applied mathematics with an application to thin airfoil theory*, SIAM Review, 10 (1968) 196-207. [7.1.3]; p. 191.

MacLachlan, James and Drake, Stillman, *Galileo's discovery of the parabolic trajectory*, Scientific American, 232 (March 1975) 102-110, 132. [7.1.0]; p. 184.

MacLane, Saunders, *Categorical algebra*, Bull. Amer. Math. Soc., 71 (1965) 40-106. [3.6.3]; p. 88.

———, *Hamiltonian mechanics and geometry*, Amer. Math. Monthly, 77 (1970) 570-586. [7.1.2]; p. 189.

———, *Modular fields*, Amer. Math. Monthly, 47 (1940) 259-274. [3.6.3]; p. 88.

———, *Some additional advances in algebra*, in A.A. Albert, Studies in Modern Algebra, M.A.A., 1963, pp. 35-58. [3.6.3]; p. 88.

———, *Some recent advances in algebra*, Amer. Math. Monthly, 46 (1939) 3-19; also in A.A. Albert, Studies in Modern Algebra, M.A.A., 1963, pp. 9-34. [3.6.2]; p. 86.

Magnus, Wilhelm and Grossman, Israel, Groups and Their Graphs, New Math. Libr., No. 14, Random House, 1964; M.A.A., 1975. [3.5.0]; p. 79.

Mahoney, Michael S., *Another look at Greek geometrical analysis*, Arch. Hist. Exact Sci., 5 (1968) 318-348. [1.2.0]; p. 5.

———, *Fermat's mathematics: proofs and conjectures*, Science, 178 (1972) 30-36. [1.2.1]; p. 9.

———, The Mathematical Career of Pierre de Fermat, Princeton, 1973. [1.3.1]; p. 19.

Maistrov, L.E., Probability Theory: A Historical Sketch, Academic Pr., 1974. [1.2.2]; p. 11.

Mal'cev, A.I., *Groups and other algebraic systems*, in A.D. Aleksandrov, et al., Mathematics--Its Content, Methods and Meaning, V. 3, M.I.T., 1963, pp. 263-351. [3.5.2]; p. 80.

Mancill, Julian D., *On the elementary transcendental functions*, in K.O. May, Lectures on Calculus, Holden-Day, 1967, pp. 15-45. [4.1.1]; p. 92.

Manes, Ernest G. and Arbib, Michael A., *Machines in a category: an expository introduction*, SIAM Review, 16 (1974) 163-192. [3.6.2]; p. 85.

Manheim, Jerome H., The Genesis of Point Set Topology, Pergamon, 1964. [1.2.2]; p. 11.

Marcus, Marvin and Minc, Henryk, *Permanents*, Amer. Math. Monthly, 72 (1965) 577-591. [3.6.1]; p. 85.

Mardzanisvili, K.K. and Postnikov, A.B., *Prime numbers*, in A.D. Aleksandrov, et al., Mathematics--Its Content, Methods and Meaning, V. 2, M.I.T., 1963, pp. 199-228. [3.3.1]; p. 68.

Marimont, Rosalind B., *Applications of graphs and Boolean matrices to computer programming*, SIAM Review, 2 (1960) 259-268. [6.5.1]; p. 179.

Marsten, R.E. and Geoffrion, A.M., *Integer programming algorithms: a framework and state-of-the-art survey*, Management Science, 18 (1972) 465-491. [6.3.3]; p. 169.

Mattson, H.F., Jr. and Assmus, E.F., Jr., *Coding and combinatorics*, SIAM Review, 16 (1974) 349-388. [7.5.3]; p. 209.

Maxfield, Margaret W. and Waugh, Frederick V., *Side-and-diagonal numbers*, Math. Magazine, 40 (1967) 74-83. [3.3.1]; p. 69.

Maxwell, E.A., Fallacies in Mathematics, Cambridge U. Pr., 1963. [1.7.1]; p. 37.

May, Robert M., *Biological populations with nonoverlapping generations: stable points, stable cycles, and chaos*, Science, 186 (1974) 645-647. [7.2.2]; p. 199.

———, *On relationships among various types of population models*, Amer. Naturalist, 107 (1973) 46-57. [7.2.2]; p. 199.

Mayer, Arthur, *Rotations and their algebra*, SIAM Review, 2 (1960) 77-122. [3.5.2]; p. 80.

Mayer, W. and Thomas, T.Y., *Foundations of the theory of Lie groups*, Annals of Math., 36 (1935) 770-822. [5.8.3]; p. 152.

McAllister, B.L., *Cyclic elements in topology: a history*, Amer. Math. Monthly, 73 (1966) 337-350. [5.5.2]; p. 142.

McHugh, James A.M., *An historical survey of ordinary linear differential equations with a large parameter and turning points*, Arch. Hist. Exact Sci., 7 (1971) 277-324. [1.2.2]; p. 11.

McKelvey, Robert, *Symmetric differential operators*, Amer. Math. Monthly, 71 (1964) 119-129. [4.6.3]; p. 119.

McLaughlin, D.W. and Keller, Joseph B., *The Feynman integral*, Amer. Math. Monthly, 82 (1975) 451-465. [4.7.3]; p. 123.

McMillan, Brockway, *An elementary approach to the theory of information*, SIAM Review, 3 (1961) 211-229. [7.5.?]; p. 209.

McShane, E.J., *A theory of limits*, in R.C. Buck, Studies in Modern Analysis, M.A.A., 1962, pp. 7-29. [4.3.2]; p. 100.

———, *A unified theory of integration*, Amer. Math. Monthly, 80 (1973) 349-359. [4.3.2]; p. 100.

———, *Partial orderings and Moore-Smith limits*, Amer. Math. Monthly, 59 (1952) 1-11. [4.3.2]; p. 101.

———, *Trends in analysis*, Amer. Math. Monthly, 74 (1967) Suppl. pp. 65-79. [4.6.2]; p. 116.

———, *Vector spaces and their applications*, in The Mathematical Sciences--A Collection of Essays for COSRIMS, M.I.T., 1969, pp. 84-96. [3.4.1]; p. 75.

Mech, William P., *Graphs of groups*, J. Undergraduate Math., 1 (1969) 97-110; 2 (1970) 37-49. [3.5.2]; p. 81.

Meisters, G.H. and Monk, J.D., *Construction of the reals via ultrapowers*, Rocky Mountain J. Math., 3 (1973) 141-158. [4.3.3]; p. 102.

Mellen, G.E., *Cryptology, computers, and common sense*, Proc. Nat. Computer Conf., 42 (1973) 569-579. [7.5.1]; p. 208.

Menger, Karl, *What is dimension?*, Amer. Math. Monthly, 50 (1943) 2-7. [5.5.1]; p. 140.

Meschkowski, Herbert, Ways of Thought of Great Mathematicians, Holden-Day, 1964. [1.3.0]; p. 18.

Meserve, Bruce E., *Decision methods for elementary algebra*, Amer. Math. Monthly, 62 (1955) 1-8. [2.5.2]; p. 53.

———, *New mathematics*, Encyclopedia Americana, 1976, V. 20, pp. 202-205. [1.6.0]; p. 33.

———, *Number systems and notation*, Encyclopedia Americana, 1976, V. 20, pp. 536f-536j. [1.2.0]; p. 5.

Metropolis, N. and Ashenhurst, R.L., *Error estimation in computer calculation*, Amer. Math. Monthly, 72 (1965) Suppl. pp. 47-58. [6.4.1]; p. 172.

Miles, John W., *Integral transforms*, in E.F. Beckenbach, Modern Mathematics for the Engineer, 2nd Ser., McGraw-Hill, 1961, pp. 68-99. [4.7.2]; p. 121.

Millman, R.S. and Stehney, Ann K., *The geometry of connections*, Amer. Math. Monthly, 80 (1973) 475-500. [5.3.3]; p. 135.

Milnor, John, *A problem in cartography*, Amer. Math. Monthly, 76 (1969) 1101-1112. [5.3.2]; p. 134.

———, *A survey of cobordism theory*, L'Enseignement Math., 8 (1962) 16-23. [5.6.3]; p. 147.

———, *Differential topology*, in T.L. Saaty, Lectures on Modern Mathematics, V. 2, Wiley, 1964, pp. 165-183. [5.6.3]; p. 147.

Minc, Henryk and Marcus, Marvin, *Permanents*, Amer. Math. Monthly, 72 (1965) 577-591. [3.6.1]; p. 85.

Minlos, R.A., *Lectures on statistical physics*, Russian Math. Surveys, 23:1 (1968) 137-196. [7.1.2]; p. 189.

Minsky, Marvin L., *Artificial intelligence*, Scientific American, 215 (September 1966) 246-260, 316; also in D. Messick, Mathematical Thinking in Behavioral Sciences, Freeman, 1968, pp. 141-148, 227. [6.5.0]; p. 178.

———, *Form and content in computer science*, J. Assoc. Comp. Mach., 17 (1970) 197-215. [6.5.0]; p. 178.

Minty, G.J., *On the axiomatic foundations of the theories of directed linear graphs, electrical networks and network-programming*, J. Math. and Mech., 15:3 (1966); also in D.R. Fulkerson, Studies in Graph Theory, Part II, M.A.A., 1975, pp. 246-300. [3.2.1]; p. 64.

Mitchell, Andrew R. and Nörlund, Niels E., *Numerical analysis*, Encyclopaedia Britannica, 15th ed., 1974, Macropaedia V. 13, pp. 381-392. [6.4.2]; p. 174.

Mityagin, B.S., *Notes on mathematical economics*, Russian Math. Surveys, 27:3 (1972) 1-19. [7.4.3]; p. 205.

Moerbeke, P.V., *Optimal stopping and free boundary problems*, Rocky Mountain J. Math., 4 (1974) 539-578. [6.1.3]; p. 159.

Molina, Edward C., *Bayes' theorem—an expository presentation*, Annals of Math. Stat., 2 (1931) 23-37. [6.1.0]; p. 154.

Monk, J. Donald, *Connections between combinatorial theory and algebraic logic*, in A. Daigneault, Studies in Algebraic Logic, M.A.A., 1974, pp. 58-91. [2.6.3]; p. 58.

———, *On the foundations of set theory*, Amer. Math. Monthly, 77 (1970) 703-711. [2.1.2]; p. 40.

——— and Meisters, G.H., *Construction of the reals via ultrapowers*, Rocky Mountain J. Math., 3 (1973) 141-158. [4.3.3]; p. 102.

Monna, A.F., Functional Analysis in Historical Perspective, Halsted Pr., 1973. [1.2.3]; p. 13.

Montgomery, David and Quirk, James, *Mathematics in economic theory*, SIAM News, 7:6 (1974) 2-3. [7.4.0]; p. 204.

Montgomery, Deane, *Oswald Veblen*, Bull. Amer. Math. Soc., 69 (1963) 26-36. [1.3.0]; p. 18.

———, *What is a topological group?*, Amer. Math. Monthly, 52 (1945) 302-307. [5.8.2]; p. 152.

Montgomery, H. and Almgren, F.J., Jr., *The 1974 Fields Medals (II): An analyst and number theorist*, Science, 186 (1974) 130-131. [1.3.3]; p. 20.

Mood, A.M., *Statistics*, McGraw-Hill Encyclopedia of Science and Technology, 1960, V. 13, pp. 66-75. [6.2.1]; p. 162.

Moore, Edward F., *Mathematics in the biological sciences*, Scientific American, 211 (September 1964) 148-164, 270; also in M. Kline, Mathematics in the Modern World, Freeman, 1968, pp. 275-283, 401-402. [7.2.0]; p. 196.

Moore, John C., *Topology*, McGraw-Hill Encyclopedia of Science and Technology, 1960, V. 13, pp. 679-683. [5.5.1]; p. 140.

Moran, P.A.P., *The probabilistic basis of stereology*, Adv. Appl. Prob., 4 (1972) Suppl. pp. 69-91. [7.6.2]; p. 212.

Mordell, L.J., *Reflections of a mathematician*, Canad. Math. Congr., 1959. [1.3.0]; p. 18.

——, *Reminiscences of an octogenarian mathematician*, Amer. Math. Monthly, 78 (1971) 952-961. [1.3.0]; p. 18.

——, Three Lectures on Fermat's Last Theorem, Macmillan, 1921; reprinted in F. Klein, et al., Famous Problems, Chelsea, 1962. [3.3.1]; p. 69.

Morgan, Bryan, Men and Discoveries in Mathematics, Transatlantic Arts, 1972. [1.2.0]; p. 5.

Moroney, M.J., Facts from Figures, Penguin, 1951. [6.2.0]; p. 161.

Morphy, Otto, *Some modern mathematical methods in the theory of lion hunting*, Amer. Math. Monthly, 75 (1968) 185-187. [1.7.2]; p. 37.

Morrey, Charles B., Jr., Schwartz, Jacob T. and Treves, François, *Functional analysis*, Encyclopaedia Britannica, 15th ed., 1974, Macropaedia V. 1, pp. 757-772. [4.6.2]; p. 117.

Morse, Marston, *Mathematics in our culture*, in T.L. Saaty and F.J. Weyl, The Spirit and Uses of the Mathematical Sciences, McGraw-Hill, 1969, pp. 105-120. [1.4.0]; p. 22.

——, *Trends in analysis*, J. Franklin Inst., 251 (1951) 33-43. [4.7.3]; p. 123.

——, *What is analysis in the large?*, Amer. Math. Monthly, 49 (1942) 358-364; also in S.S. Chern, Studies in Global Geometry and Analysis, M.A.A., 1967, pp. 5-15. [5.6.2]; p. 145.

Morse, Philip M., *Mathematical problems in operations research*, Bull. Amer. Math. Soc., 54 (1948) 602-621. [6.3.2]; p. 169.

Moser, Leo and Butchart, J.H., *No calculus, please*, Scripta Math., 18 (1952) 221-236. [4.1.1]; p. 91.

—— and Harary, Frank, *The theory of round robin tournaments*, Amer. Math. Monthly, 73 (1966) 231-246. [3.1.2]; p. 61.

Mosteller, Frederick and Hoaglin, David C., *Statistics*, Encyclopaedia Britannica, 15th ed., 1974, Macropaedia V. 17, pp. 615-624. [6.2.1]; p. 162.

Mostow, George D. and Thom, René F., *Topological groups and differential topology*, Encyclopaedia Britannica, 15th ed., 1974, Macropaedia, V. 18, pp. 489-504. [5.8.3]; p. 152.

Mostowski, Andrezej, Thirty Years of Foundational Studies, Barnes & Noble, 1966. [1.2.2]; p. 11.

Munroe, M.E., *Bringing calculus up to date*, Amer. Math. Monthly, 65 (1958) 81-90. [4.3.2]; p. 101.

———, *Mathematical theory of probability*, Encyclopedia Americana, 1976, V. 22, pp. 622-625. [6.1.1]; p. 156.

Murnaghan, F.D., *An elementary presentation of the theory of quaternions*, Scripta Math., 10 (1944) 37-49. [3.4.1]; p. 75.

Murray, Francis J. and Ford, Lester R., *Mathematics as a calculatory science*, Encyclopaedia Britannica, 15th ed., 1974, Macropaedia V. 11, pp. 671-696. [1.5.0]; p. 27.

Murray, J.D., *Approximate methods in mathematics*, Math. Spectrum, 6 (1973-74) 19-24. [4.2.2]; p. 97.

Myhill, John, *What is a real number?*, Amer. Math. Monthly, 79 (1972) 748-754. [2.3.1]; p. 47.

Nagel, Ernest, *The meaning of probability*, J. Amer. Stat. Assoc., 31 (1936) 10-30; also in J.R. Newman, The World of Mathematics, V. 2, Simon and Schuster, 1956, pp. 1398-1414. [6.1.0]; p. 155.

——— and Newman, James R., *Gödel's proof*, Scientific American, 194 (June 1956) 71-86, 168, 170; also in M. Kline, Mathematics in the Modern World, Freeman, 1968, pp. 221-230, 400; in J.R. Newman, The World of Mathematics, V. 3, 1956, pp. 1668-1695; also published by New York U. Pr., 1958. [2.5.0]; p. 52.

Nalimov, V.V., *Logical foundations of applied mathematics*, Synthese, 27 (1974) 211-250. [1.5.2]; p. 30.

Naps, Thomas L., *Arithmetical formalism--Hilbert's proof theory and Gödel's proof*, J. Undergraduate Math., 1 (1969) 111-132. [2.5.1]; p. 53.

Narasimhan, Raghavan and Iwasawa, Kenkichi, *Complex analysis*, Encyclopaedia Britannica, 15th ed., 1974, Macropaedia V. 1, pp. 719-735. [4.4.2]; p. 105.

Nash, John, *Non-cooperative games*, Annals of Math., 54 (1951) 286-295. [6.3.2]; p. 169.

———, *Two-person cooperative games*, Econometrica, 21 (1953) 128-140. [6.3.2]; p. 169.

Nash-Williams, C. St.J.A., *Hamiltonian circuits*, in D.R. Fulkerson, Studies in Graph Theory, Part II, M.A.A., 1975, pp. 301-360. [3.2.3]; p. 65.

Nashed, M.Z., *Some remarks on variations and differentials*, Amer. Math. Monthly, 73 (1966) Suppl. pp. 63-76. [4.3.2]; p. 101.

Nath, Prem and Varma, R.S., *Information theory--a survey*, J. Math. Sci., 2 (1967) 75-109. [7.5.2]; p. 209.

Nelson, R.J. and Rankin, Bayard, *Automata theory*, Encyclopaedia Britannica, 15th ed., 1974, Macropaedia V. 2, pp. 497-505. [6.5.1]; p. 179.

Neményi, P.F., *The main concepts and ideas of fluid dynamics in their historical development*, Arch. Hist. Exact Sci., 2 (1962) 52-86. [1.2.0]; p. 5.

Neuts, Marcel F., *Are many 1-1 functions on the positive integers onto?*, Math. Magazine, 41 (1968) 103-109. [6.1.2]; p. 158.

Nevanlinna, Rolf, *Methods in the theory of integral and meromorphic functions*, J. London Math. Soc., 41 (1966) 11-28. [4.4.2]; p. 106.

────, *Reform in teaching mathematics*, Amer. Math. Monthly, 73 (1966) 451-464. [1.6.0]; p. 33.

Newman, D.J. and Klamkin, Murray S., *The philosophy and applications of transform theory*, SIAM Review, 3 (1961) 10-36. [7.1.2]; p. 189.

Newman, James R. and Kasner, Edward, *Paradox lost and paradox regained*, in J.R. Newman, The World of Mathematics, V. 3, Simon and Schuster, 1956, pp. 1936-1955. [2.1.0]; p. 38.

──── and Nagel, Ernest, *Gödel's proof*, Scientific American, 194 (June 1956) 71-86, 168, 170; also in M. Kline, Mathematics in the Modern World, Freeman, 1968, pp. 221-230, 400; in J.R. Newman, The World of Mathematics, V. 3, 1956, pp. 1668-1695; also published by New York U. Pr., 1958. [2.5.0]; p. 52.

Newman, M.H.A., *What is mathematics? New answers to an old question*, Math. Gazette, 43 (1959) 161-171. [1.5.0]; p. 27.

Newns, W.F., *Functional dependence*, Amer. Math. Monthly, 74 (1967) 911-920. [4.3.2]; p. 101.

Nikol'skiĭ, S.M., *Approximations of functions*, in A.D. Aleksandrov, et al., Mathematics--Its Content, Methods and Meaning, V. 2, M.I.T., 1963, pp. 265-302. [6.4.1]; p. 173.

──── and Lavrent'ev, M.A., *Analysis*, in A.D. Aleksandrov, et al., Mathematics--Its Content, Methods and Meaning, V. 1, M.I.T., 1963, pp. 65-180. [4.1.1]; p. 92.

Nirenberg, L., *Partial differential equations with applications in geometry*, in T.L. Saaty, Lectures on Modern Mathematics, V. 2, Wiley, 1964, pp. 1-41. [4.5.2]; p. 110.

Nitsche, Johannes C.C., *Plateau's problems and their modern ramifications*, Amer. Math. Monthly, 81 (1974) 945-968. [5.3.2]; p. 134.

Niven, Ivan M., *Formal power series*, Amer. Math. Monthly, 76 (1969) 871-889. [4.2.2]; p. 97.

────, *Mathematics of Choice--or How to Count Without Counting*, New Math. Libr., No. 15, Random House, 1965; M.A.A., 1975. [3.1.1]; p. 60.

———, *Numbers: Rational and Irrational*, New Math. Libr., No. 1, Random House, 1961; M.A.A., 1975. [3.3.0]; p. 68.

——— and Zuckerman, H.S., *Lattice points and polygonal area*, Amer. Math. Monthly, 74 (1967) 1195-1200. [5.1.1]; p. 127.

Nomizu, Katsumi, *Recent developments in the theory of connections and holonomy groups*, Advances in Math., 1 (1965) 1-49. [5.3.3]; p. 135.

Nörlund, Niels E. and Mitchell, Andrew R., *Numerical analysis*, Encyclopaedia Britannica, 15th ed., 1974, Macropaedia V. 13, pp. 381-392. [6.4.2]; p. 174.

Novikov, S.P., *The main trends of algebraic topology and algebraic geometry*, Russian Math. Surveys, 19:6 (1964) 67-74. [5.7.3]; p. 150.

Nový, Luboš, The Origins of Modern Algebra, Noordhoff, 1974. [1.2.2]; p. 12.

Oettinger, A.G., *Computational linguistics*, Amer. Math. Monthly, 72 (1965) Suppl. pp. 147-150. [7.6.2]; p. 212.

Olds, Carl Douglas, Continued Fractions, New Math. Libr., No. 9, Random House, 1963; M.A.A., 1975. [3.3.0]; p. 68.

———, *The simple continued fraction expansion of e*, Amer. Math. Monthly, 77 (1970) 968-974. [3.3.2]; p. 71.

O'Neil, P.V., *Ulam's conjecture and graph reconstructions*, Amer. Math. Monthly, 77 (1970) 35-43. [3.2.1]; p. 64.

Ordman, Edward T., *One and one is nothing: liberating mathematics*, Soundings, 56 (1973) 164-181. [1.6.0]; p. 33.

Ore, Oystein, Graphs and Their Uses, New Math. Libr., No. 10, Random House, 1963; M.A.A., 1975. [3.2.0]; p. 63.

———, Invitation to Number Theory, New Math. Libr., No. 20, Random House, 1967; M.A.A., 1975. [3.3.1]; p. 69.

———, Niels Henrik Abel, Mathematician Extraordinary, U. Minn. Pr., 1957; Chelsea, 1974. [1.3.0]; p. 18.

———, *Pascal and the invention of probability theory*, Amer. Math. Monthly, 67 (1960) 409-419. [6.1.0]; p. 155.

Ornstein, D.S., *Measure-preserving transformations and random processes*, Amer. Math. Monthly, 78 (1971) 833-840. [4.7.3]; p. 123.

Oxnard, Charles E., *Mathematics, shape and function: a study in primate anatomy*, Amer. Scientist, 57 (1969) 75-96. [7.2.1]; p. 197.

Oxtoby, John C., *What are physical dimensions?*, Amer. Physics Teacher, 2 (Sept. 1934) 85-90. [7.1.0]; p. 185.

Packel, Edward W., *Hilbert space operators and quantum mechanics*, Amer. Math. Monthly, 81 (1974) 863-873. [4.6.2]; p. 117.

Page, David A., *Probability*, NCTM Twenty-Fourth Yearbook, 1959, 229-271. [6.1.0]; p. 155.

Paige, Lowell J., *Jordan algebras*, in A.A. Albert, Studies in Modern Algebra, M.A.A., 1963, pp. 144-186. [3.4.2]; p. 76.

Panati, Charles, *Catastrophe theory*, Newsweek (January 19, 1976) 54-55. [7.3.0]; p. 201.

Papanicolaou, G.C., *Stochastic equations and their applications*, Amer. Math. Monthly, 80 (1973) 526-545. [4.5.3]; p. 112.

Parlett, Beresford, *Matrix eigenvalue problems*, Amer. Math. Monthly, 72 (1965) Suppl. pp. 59-66. [6.4.1]; p. 173.

Passman, D.S., *What is a group ring?*, Amer. Math. Monthly, 83 (1976) 173-185. [3.5.2]; p. 81.

Payne, Lawrence F., *Isoperimetric inequalities and their applications*, SIAM Review, 9 (1967) 453-488. [6.3.3]; p. 169.

Pedoe, Daniel, *On a theorem in geometry*, Amer. Math. Monthly, 74 (1967) 627-640. [5.1.2]; p. 128.

————, The Gentle Art of Mathematics, Macmillan, 1958. [1.1.0]; p. 1.

Pekeris, C.L., *Adventures in applied mathematics*, Quarterly of Applied Math., 30 (1972) 67-83. [7.1.2]; p. 189.

Perlis, Alan J., *Automatic programming*, Quarterly of Applied Math., 30 (1972) 85-90. [6.5.0]; p. 178.

Pétard, H., *A contribution to the mathematical theory of big game hunting*, Amer. Math. Monthly, 45 (1938) 446-447. [1.7.2]; p. 37.

Peterson, Elmer L., *Geometric programming*, SIAM Review, 18 (1976) 1-51. [6.3.3]; p. 170.

Peterson, W. Wesley, *Error-correcting codes*, Scientific American, 206 (February 1962) 96-108, 188; also in D. Messick, Mathematical Thinking in Behavioral Sciences, Freeman, 1968, pp. 52-58, 224. [7.5.0]; p. 208.

Petrovskiĭ, I.G., *Ordinary differential equations*, in A.D. Aleksandrov, et al., Mathematics--Its Content, Methods and Meaning, V. 1, M.I.T., 1963, pp. 311-356. [4.5.1]; p. 109.

Phillips, Anthony, *Topology*, Encyclopedia Americana, 1976, V. 26, pp. 850-854. [5.5.1]; p. 140.

————, *Turning a surface inside out*, Scientific American, 214 (May 1966) 112-120, 148. [5.6.1]; p. 145.

Phillips, Keith L., *The maximal theorems of Hardy and Littlewood*, Amer. Math. Monthly, 74 (1967) 648-660. [4.3.3]; p. 102.

Phillips, Ralph S., *Semigroup methods in the theory of partial differential equations*, in E.F. Beckenbach, Modern Mathematics for the Engineer, 2nd Ser., McGraw-Hill, 1961, pp. 100-132. [4.5.3]; p. 112.

Piaget, Jean, Genetic Epistemology, Columbia U. Pr., 1970. [1.6.0]; p. 33.

Pierpont, James, Mathematical rigor, past and present, Bull. Amer. Math. Soc., 34 (1928) 23-53. [1.2.1]; p. 9.

──────, The history of mathematics in the nineteenth century, Bull. Amer. Math. Soc., 11 (1904) 136-159. [1.2.2]; p. 12.

Poénaru, Valentin, On the geometry of differentiable manifolds, in P.J. Hilton, Studies in Modern Topology, M.A.A., 1968, pp. 165-207. [5.6.3]; p. 147.

Pogorzelski, H.A. and Davis, Chandler, Contemporary mathematical notation, McGraw-Hill Encyclopedia of Science and Technology, 1960, V. 8, pp. 172-174. [1.2.1]; p. 7.

Poincaré, Henri, Mathematical creation, Scientific American, 179 (August 1948) 54-57; also in M. Kline, Mathematics in the Modern World, Freeman, 1968, pp. 14-17; and in J.R. Newman, The World of Mathematics, V. 4, Simon and Schuster, 1956, pp. 2041-2050. [1.5.0]; p. 27.

Polachek, Harry, The structure of the honeycomb, Scripta Math., 7 (1940) 87-98. [7.2.1]; p. 197.

Polak, E., An historical survey of computational methods in optimal control, SIAM Review, 15 (1973) 553-584. [6.3.3]; p. 170.

Pollak, Henry O., How can we teach applications of math?, Educ. Studies Math., 2 (1969-70) 393-404. [1.6.1]; p. 34.

──────, On some of the problems of teaching applications of mathematics, Educ. Studies Math., 1 (1968-69) 24-30. [1.6.1]; p. 34.

Pollock, John L., Mathematical proof, Amer. Phil. Quarterly, 4 (1967) 238-244. [2.2.0]; p. 43.

Pólya, George, Circle, sphere, symmetrization and some classical physical problems, in E.F. Beckenbach, Modern Mathematics for the Engineer, 2nd Ser., McGraw-Hill, 1961, pp. 420-441. [4.7.2]; p. 121.

──────, Heuristic reasoning in the theory of numbers, Amer. Math. Monthly, 66 (1959) 375-384. [3.3.1]; p. 69.

──────, How to Solve It, Princeton U. Pr., 1945; excerpted in J.R. Newman, The World of Mathematics, V. 3, Simon and Schuster, 1956, pp. 1980-1992. [1.6.0]; p. 33.

──────, Mathematical Discovery, Wiley, V. 1, 1962, V. 2, 1965. [1.6.1]; p. 34.

──────, Mathematics and Plausible Reasoning: V. I, Induction and Analogy in Mathematics; V. II, Patterns of Plausible Inference, Princeton U. Pr., 1954. [1.6.1]; p. 34.

Postnikov, A.B. and Mardzanisvili, K.K., Prime numbers, in A.D. Aleksandrov, et al., Mathematics--Its Content, Methods and Meaning, V. 2, M.I.T., 1963, pp. 199-228. [3.3.1]; p. 68.

Posy, Carl J., *Brouwer's constructivism*, Synthese, 27 (1974) 125-159. [2.3.2]; p. 48.

Powell, M.J.D., *A survey of numerical methods for unconstrained optimization*, SIAM Review, 12 (1970) 79-97. [6.3.2]; p. 169.

Prager, W., *Mathematical programming and theory of structures*, SIAM J. Appl. Math., 13 (1965) 312-332. [6.3.3]; p. 170.

Pratt, John W., Raiffa, Howard and Schlaifer, Robert, *The foundations of decision under uncertainty: an elementary exposition*, J. Amer. Stat. Assoc., 59 (1964) 353-375. [6.2.2]; p. 162.

Prenowitz, Walter, *A contemporary approach to classical geometry*, Amer. Math. Monthly, 68 (1961) Suppl. pp. 1-67. [5.1.1]; p. 127.

Priest, Graham, *A bedside reader's guide to the conventionalist philosophy of mathematics*, in J. Bell, et al., Proc. Bertrand Russell Memorial Logic Conference, Leeds, 1973, pp. 115-132. [2.2.0]; p. 43.

Protter, M.H., *Potentials*, McGraw-Hill Encyclopedia of Science and Technology, 1960, V. 10, pp. 537-539. [4.2.1]; p. 95.

Putnam, Hilary, *Mathematics without foundations*, J. Phil., 64 (1967) 5-22. [2.2.0]; p. 44.

———, *Recursive functions and hierarchies*, Amer. Math. Monthly, 80 (1973) Suppl. pp. 68-86. [2.6.2]; p. 56.

Quine, Willard Van Orman, *Paradox*, Scientific American, 206 (April 1962) 84-96, 193; also in M. Kline, Mathematics in the Modern World, Freeman, 1968, pp. 200-208, 399. [2.6.0]; p. 55.

———, *The foundations of mathematics*, Scientific American, 211 (September 1964) 112-127, 269; also in M. Kline, Mathematics in the Modern World, Freeman, 1968, pp. 191-199, 399. [2.2.0]; p. 44.

Quirk, James and Montgomery, David, *Mathematics in economic theory*, SIAM News, 7:6 (1974) 2-3. [7.4.0]; p. 204.

Rabinowitz, Philip, *Applications of linear programming to numerical analysis*, SIAM Review, 10 (1968) 121-159. [6.4.2]; p. 174.

Rademacher, Hans, *Number theory*, McGraw-Hill Encyclopedia of Science and Technology, 1960, V. 9, pp. 224-227. [3.3.1]; p. 69.

———, *Trends in research: the analytic number theory*, Bull. Amer. Math. Soc., 48 (1942) 379-401. [3.3.3]; p. 73.

——— and Toeplitz, Otto, The Enjoyment of Mathematics: Selections from Mathematics for the Amateur, Princeton U. Pr., 1957. [1.1.1]; p. 2.

Radó, Tibor, *What is the area of a surface?*, Amer. Math. Monthly, 50 (1943) 139-141. [4.2.1]; p. 95.

Raiffa, Howard, Pratt, John W. and Schlaifer, Robert, *The foundations of decision under uncertainty: an elementary exposition*, J. Amer. Stat. Assoc., 59 (1964) 353-375. [6.2.2]; p. 162.

Randall, C.H. and Foulis, D.J., *An approach to empirical logic*, Amer. Math. Monthly, 77 (1970) 363-374. [2.6.1]; p. 55.

Randell, Brian, The Origin of Digital Computers, Springer-Verlag, 1973. [6.5.0]; p. 178.

Rankin, Bayard and Nelson, R.J., *Automata theory*, Encyclopaedia Britannica, 15th ed., 1974, Macropaedia V. 2, pp. 497-505. [6.5.1]; p. 179.

Rapoport, Anatol, *Critiques of game theory*, Behavioral Science, 4 (1959) 49-66. [6.3.0]; p. 165.

———, *Directions in mathematical psychology*, Amer. Math. Monthly, 83 (1976) 85-106, 153-172. [7.3.2]; p. 202.

———, *Escape from paradox*, Scientific American, 217 (July 1967) 50-56, 134. [6.3.0]; p. 165.

———, *The use and misuse of game theory*, Scientific American, 207 (December 1962) 108-118, 192; also in M. Kline, Mathematics in the Modern World, Freeman, 1968, pp. 304-312, 402-403; and in D. Messick, Mathematical Thinking in Behavioral Sciences, Freeman, 1968, pp. 95-103, 226. [6.3.0]; p. 165.

———, *Uses of mathematics outside the physical sciences*, SIAM Review, 15 (1973) 481-502. [7.3.1]; p. 202.

——— and Rebhun, L.I., *On the mathematical theory of rumor spread*, Bull. Math. Biophys., 14 (1952) 375-383. [7.3.2]; p. 202.

Rashevsky, N., *Topology and life--in search of general mathematical principles in biology and sociology*, Bull. Math. Biophys., 16 (1954) 317-348. [7.2.1]; p. 197.

Rasiowa, Helena, *Post algebras as a semantic foundation of m-valued logics*, in A. Daigneault, Studies in Algebraic Logic, M.A.A., 1974, pp. 92-142. [2.6.2]; p. 56.

Ravetz, Jerome R., *Fourier series*, Encyclopedia Americana, 1976, V. 11, pp. 657-658. [4.2.1]; p. 95.

Rebhun, L.I. and Rapoport, Anatol, *On the mathematical theory of rumor spread*, Bull. Math. Biophys., 14 (1952) 375-383. [7.3.2]; p. 202.

Redheffer, Raymond, *The homotopy theorems of function theory*, Amer. Math. Monthly, 76 (1969) 778-787. [4.4.2]; p. 106.

——— and Johnson, Paul, *Scrambled series*, Amer. Math. Monthly, 73 (1966) 822-828. [4.2.2]; p. 97.

Rees, Mina and Shenton, Walter F., *Algebra*, Encyclopedia Americana, 1976, V. 1, pp. 555-562. [3.6.1]; p. 85.

Reichmann, W.J., Use and Abuse of Statistics, Penguin, 1964. [6.2.0]; p. 161.

Reid, Constance, Hilbert, Springer-Verlag, 1970. [1.3.0]; p. 18.

Reid, W.T., *Anatomy of the ordinary differential equation*, Amer. Math. Monthly, 82 (1975) 971-984. [4.5.2]; p. 110.

Rényi, Alfréd, *A Socratic dialogue on mathematics*, Canad. Math. Bull., 7 (1964) 441-462; also in A. Rényi, Dialogues on Mathematics, Holden-Day, 1967, pp. 3-25. [1.5.0]; p. 28.

Reyes, Gonzalo E., *From sheaves to logic*, in A. Daigneault, Studies in Algebraic Logic, M.A.A., 1974, pp. 143-204. [2.6.3]; p. 58.

Reza, Fazlollah, *Information theory*, Encyclopedia Americana, 1976, V. 15, pp. 166-168. [7.5.1]; p. 208.

Rhodes, F., *$1 - 1 + 1 - 1 + \ldots = 1/2$?*, Math. Gazette, 55 (1971) 298-305. [4.2.1]; p. 95.

Richards, Ian, *Impossibility*, Math. Magazine, 48 (1975) 249-262. [3.6.1]; p. 85.

Richmond, D.E., *Areas and volumes without limit processes*, Amer. Math. Monthly, 73 (1966) 477-483. [4.1.1]; p. 92.

Rindler, W., *Survey of relativity theory*, SIAM Review, 3 (1961) 105-118. [7.1.1]; p. 186.

Robbins, Herbert, *Optimal stopping*, Amer. Math. Monthly, 77 (1970) 333-343. [6.1.2]; p. 158.

——, *The theory of probability*, NCTM Twenty-Third Yearbook, 1957, 336-371. [6.1.2]; p. 158.

—— and Courant, Richard, *Topology*, in R. Courant and H. Robbins, What is Mathematics, Oxford U. Pr., 1941, pp. 235-271; also in J. R. Newman, The World of Mathematics, V. 1, Simon and Schuster, 1956, pp. 581-599. [5.5.0]; p. 139.

—— and ——, *What is Mathematics?*, Oxford U. Pr., 1941. [1.1.1]; p. 2.

Roberts Fred S. and Brown, Thomas A., *Signed digraphs and the energy crisis*, Amer. Math. Monthly, 82 (1975) 577-594. [7.6.1]; p. 211.

Robinson, Abraham, *Between logic and mathematics*, I.C.S.U. Rev. World Sci., 6 (1964) 218-226. [2.6.1]; p. 56.

——, *Formalism 64*, in Y. Bar-Hillel, Logic, Methodology and Philosophy of Science, North-Holland, 1965, pp. 228-248. [2.2.0]; p. 44.

——, *From a formalist's point of view*, Dialectica, 23 (1969) 45-49. [2.2.0]; p. 44.

——, *Function theory on some nonarchimedian fields*, Amer. Math. Monthly, 80 (1973) Suppl. pp. 87-109. [2.4.2]; p. 50.

——, *Model theory as a framework for algebra*, in M.D. Morley, Studies in Model Theory, M.A.A., 1973, pp. 134-157. [2.6.2]; p. 57.

——, *Numbers—what are they and what are they good for?*, Yale Scientific, (May 1973) 14-16. [2.4.0]; p. 49.

——, *Some thoughts on the history of mathematics*, Compositio Math., 20 (1968) 188-193. [2.4.1]; p. 49.

——, *Standard and nonstandard number systems*, Nieuw Archief Wiskunde, 21 (1973) 115-133. [2.4.2]; p. 50.

——, *The metaphysics of the calculus*, in J. Hintikka, The Philosophy of Mathematics, Oxford U. Pr., 1969, pp. 153-163; also in I. Lakatos, Problems in the Philosophy of Mathematics, North-Holland, 1967, pp. 28-40. [2.4.1]; p. 49.

Robinson, J.A., *Theorem-proving on the computer*, J. Assoc. Comp. Mach., 10 (1963) 163-174. [6.5.2]; p. 180.

Robinson, Julia, *Diophantine decision problems*, in W.J. LeVeque, Studies in Number Theory, M.A.A., 1969, pp. 76-116. [3.3.3]; p. 73.

Robinson, Louis, et al., *Computers*, Encyclopedia Americana, 1976, V. 7, pp. 472-494. [6.5.0]; p. 178.

Robinson, Raphael M., *The converse of Fermat's theorem*, Amer. Math. Monthly, 64 (1957) 703-710. [3.3.2]; p. 71.

Rogers, Hartley R., Jr., *An example in mathematical logic*, Amer. Math. Monthly, 70 (1963) 929-945. [2.6.1]; p. 56.

——, *Information theory*, Math. Magazine, 37 (1964) 63-78. [7.5.1]; p. 208.

——, *The present state of Turing machine computability*, SIAM J. Appl. Math., 7 (1959) 114-130. [6.5.2]; p. 180.

Rogers, Pat, *The parallel axiom*, Math. Spectrum, 5 (1972-73) 58-66. [5.1.0]; p. 126.

Rogers, Robert, *Mathematical and philosophical analyses*, Phil. Sci., 31 (1964) 255-264. [2.2.0]; p. 44.

Rosen, Robert, *On mathematics and biology*, in T.L. Saaty and F.J. Weyl, The Spirit and Uses of the Mathematical Sciences, McGraw-Hill, 1969, pp. 203-218. [7.2.0]; p. 196.

Rosen, Saul, *Electronic computers: a historical survey*, Computing Surveys, 1 (1969) 7-36. [1.2.0]; p. 5.

Rosenkrantz, R.D., *The significance test controversy*, Synthese, 26 (1973) 304-321. [6.2.2]; p. 163.

Rosenlicht, Maxwell, *Integration in finite terms*, Amer. Math. Monthly, 79 (1972) 963-972. [4.1.2]; p. 92.

Rosenthal, Arthur, *The history of calculus*, Amer. Math. Monthly, 58 (1951) 75-86. [1.2.1]; p. 9.

——, *What are set functions?*, Amer. Math. Monthly, 55 (1948) 14-20. [4.3.2]; p. 101.

Rosser, J. Barkley, *An informal exposition of proofs of Gödel's theorems and Church's theorem*, J. Symbolic Logic, 4 (1939) 53-60. [2.5.2]; p. 53.

─────, *Asymptotic formulas and series*, in E.F. Beckenbach, <u>Modern Mathematics for the Engineer</u>, 2nd Ser., McGraw-Hill, 1961, pp. 133-163. [4.7.3]; p. 123.

Rota, Gian-Carlo, *Combinatorial analysis*, in <u>The Mathematical Sciences--A Collection of Essays for COSRIMS</u>, M.I.T., 1969, pp. 197-208. [3.1.1]; p. 61.

───── and Harper, L.H., *Matching theory: an introduction*, <u>Adv. in Prob.</u>, 1 (1971) 169-215. [3.1.2]; p. 61.

Royden, H.L., *Function algebras*, <u>Bull. Amer. Math. Soc.</u>, 69 (1963) 281-298. [4.6.3]; p. 119.

Rubin, Jean E., *Finite sets*, <u>Math. Magazine</u>, 46 (1973) 183-192. [2.1.2]; p. 40.

─────, *The compactness theorem in mathematical logic*, <u>Math. Magazine</u>, 46 (1973) 261-265. [2.6.2]; p. 57.

Rudin, Mary Ellen, *Souslin's conjecture*, <u>Amer. Math. Monthly</u>, 76 (1969) 1113-1119. [5.5.2]; p. 142.

Saaty, Thomas L., *Operations research: some contributions to mathematics*, <u>Science</u>, 178 (1972) 1061-1070. [6.3.0]; p. 165.

─────, *Remarks on the four color problem: the Kempe catastrophe*, <u>Math. Magazine</u>, 40 (1967) 31-36. [3.2.0]; p. 63.

─────, *Thirteen colorful variations on Guthrie's four-color conjecture*, <u>Amer. Math. Monthly</u>, 79 (1972) 2-43. [3.2.2]; p. 64.

───── and Alexander, Joyce M., *Optimization and the geometry of numbers: packing and covering*, <u>SIAM Review</u>, 17 (1975) 475-519. [3.3.1]; p. 69.

Sadovskii, L.E. and Lifshits, V.N., *Algebraic models of computing machines*, <u>Russian Math. Surveys</u>, 27:3 (1972) 87-135. [6.5.3]; p. 181.

Sagan, Hans, *Area and integration*, in K.O. May, <u>Lectures on Calculus</u>, Holden-Day, 1967, pp. 61-72. [4.2.1]; p. 95.

Salam, Abdus, *Theory of groups and the symmetry physicist*, <u>J. London Math. Soc.</u>, 41 (1966) 49-62. [5.8.3]; p. 153.

Salkind, Charles T., <u>The Contest Problem Book</u>, New Math. Libr., No. 5, Random House, 1961; M.A.A., 1975. [1.7.1]; p. 37.

─────, <u>The M.A.A. Problem Book II</u>, New Math. Libr., No. 17, Random House, 1966: M.A.A., 1975. [1.7.1]; p. 37.

───── and Earl, James M., <u>The M.A.A. Problem Book III</u>, New Math. Libr., No. 25, Random House, 1973; M.A.A., 1975. [1.7.1]; p. 37.

Salmon, Wesley, C., *Confirmation*, <u>Scientific American</u>, 228 (May 1973) 75-83, 120. [1.5.0]; p. 28.

Samuel, Pierre, *Unique factorization*, <u>Amer. Math. Monthly</u>, 75 (1968) 945-952. [3.6.3]; p. 88.

Samuelson, Paul A., *Mathematics of speculative price*, SIAM Review, 15 (1973) 1-42. [7.4.2]; p. 205.

———, *Maximum principles in analytical economics*, Science, 173 (1971) 991-997. [7.4.1]; p. 204.

Savage, Leonard J., *The foundations of statistics reconsidered*, in H.E. Kyburg, Jr. and H.E. Smokler, Studies in Subjective Probability, Wiley, 1964, pp. 173-188. [6.2.1]; p. 162.

———, *The theory of statistical decisions*, J. Amer. Stat. Assoc., 46 (1951) 55-67. [6.2.2]; p. 163.

Sawyer, W.W., *A reflection on foundations of mathematics--mathematicians regarded as biological specimens*, Philosophia Mathematica, 1 (1964) 5-32. [2.2.0]; p. 44.

———, *What is Calculus About?*, New Math. Libr., No. 2, Random House, 1961; M.A.A., 1975. [4.1.1]; p. 92.

Schaefer, H.H., *A brief introduction to the Lebesgue-Stieltjes integral*, in I.I. Hirschman, Jr., Studies in Real and Complex Analysis, M.A.A., 1965, pp. 90-123. [4.3.2]; p. 101.

Scherk, Peter, *Some concepts of conformal geometry*, Amer. Math. Monthly, 67 (1960) 1-30. [5.1.2]; p. 128.

Schiffer, Menahem M., *Boundary value problems in elliptic partial differential equations*, in E.F. Beckenbach, Modern Mathematics for the Engineer, McGraw-Hill, 1956, pp. 110-144. [4.5.2]; p. 110.

Schild, Alfred, *The clock paradox in relativity theory*, Amer. Math. Monthly, 66 (1959) 1-18. [7.1.1]; p. 186.

Schlaifer, Robert, *Expected value and utility*, in B. Lieberman, Contemporary Problems in Statistics, Ocford U. Pr., 1971, pp. 250-266. [7.4.0]; p. 204.

———, *The meaning of probability*, in B. Lieberman, Contemporary Problems in Statistics, Oxford U. Pr., 1971, pp. 236-249. [6.1.0]; p. 155.

———, Pratt, John W. and Raiffa, Howard, *The foundations of decision under uncertainty: an elementary exposition*, J. Amer. Stat. Assoc., 59 (1964) 353-375. [6.2.2]; p. 162.

Schmidt, Wolfgang M., *Approximation to algebraic numbers*, L'Enseignement Math., 17 (1971) 187-253. [3.3.2]; p. 71.

Schoenberg, I.J., *The elementary cases of Landau's problem of inequalities between derivatives*, Amer. Math. Monthly, 80 (1973) 121-158. [6.4.2]; p. 174.

Schultz, Reinhard, *Some recent results on topological manifolds*, Amer. Math. Monthly, 78 (1971) 941-952. [5.7.3]; p. 150.

Schwartz, Jacob T., *Functional analysis*, in The Mathematical Sciences--A Collection of Essays for COSRIMS, M.I.T., 1969, pp. 72-83. [4.6.1]; p. 115.

———, *Semantic and syntactic issues in programming*, Bull. Amer. Math. Soc., 80 (1974) 185-206. [6.5.2]; p. 180.

———, *The pernicious influence of mathematics on science*, in E. Nagel, P. Suppes and A. Tarski, <u>Logic, Methodology and Philosophy of Science</u>, Stanford, 1962, pp. 356-360. [1.4.1]; p. 23.

———, Morrey, Charles B., Jr. and Treves, François, *Functional analysis*, <u>Encyclopaedia Britannica</u>, 15th ed., 1974, Macropaedia V. 1, pp. 757-772. [4.6.2]; p. 117.

Schwartz, Laurent, *Some applications of the theory of distributions*, in T.L. Saaty, <u>Lectures on Modern Mathematics</u>, V. 1, Wiley, 1963, pp. 23-58. [4.6.3]; p. 119.

Scott, Dana, *A proof of the independence of the continuum hypothesis*, <u>Math. Systems Theory</u>, 1 (1967) 89-111. [2.1.2]; p. 40.

Searle, John, *Chomsky's revolution in linguistics*, <u>N.Y. Review of Books</u>, 18 (June 29, 1972) 16-24. [7.6.0]; p. 211.

Seebach, J. Arthur, Jr., Seebach, Linda A. and Steen, Lynn Arthur, *What is a sheaf?*, <u>Amer. Math. Monthly</u>, 77 (1970) 681-703. [5.6.2]; p. 146.

Seebach, Linda A., Seebach, J. Arthur, Jr. and Steen, Lynn Arthur, *What is a sheaf?*, <u>Amer. Math. Monthly</u>, 77 (1970) 681-703. [5.6.2]; p. 146.

Seeley, Robert T., *Spherical harmonics*, <u>Amer. Math. Monthly</u>, 73 (1966) Suppl. pp. 115-121. [4.7.3]; p. 124.

Segal, Irving, *Algebraic integration theory*, <u>Bull. Amer. Math. Soc.</u>, 71 (1965) 419-489. [4.3.3]; p. 102.

Segel, Lee A., *Simplification and scaling*, <u>SIAM Review</u>, 14 (1972) 547-571. [7.6.1]; p. 211.

———, *The importance of asymptotic analysis in applied mathematics*, <u>Amer. Math. Monthly</u>, 73 (1966) 7-14. [4.7.3]; p. 124.

Seidenberg, A., *On the area of a semi-circle*, <u>Arch. Hist. Exact Sci.</u>, 9 (1972) 171-211. [1.2.0]; p. 5.

Seifert, Herbert and Threlfall, William, *Old and new results on knots*, <u>Canad. J. Math.</u>, 2 (1950) 1-15. [5.7.3]; p. 150.

Serrin, James, *On the area of curved surfaces*, <u>Amer. Math. Monthly</u>, 68 (1961) 435-440. [4.3.2]; p. 101.

Shapley, L.S. and Gale, David, *College admissions and the stability of marriage*, <u>Amer. Math. Monthly</u>, 69 (1962) 9-15. [6.3.0]; p. 165.

Shenton, Walter F., *Mathematical signs and symbols*, <u>Encyclopedia Americana</u>, 1976, V. 18, pp. 426-428. [1.2.1]; p. 9.

——— and Rees, Mina, *Algebra*, <u>Encyclopedia Americana</u>, 1976, V. 1, pp. 555-562. [3.6.1]; p. 85.

Shephard, G.C. and Grünbaum, Branko, *Convex polytopes*, <u>Bull. London Math. Soc.</u>, 1 (1969) 257-300. [5.2.3]; p. 132.

Shepherdson, J.C. and Sturgis, H.E., *Computability of recursive functions*, <u>J. Assoc. Comp. Mach.</u>, 10 (1963) 217-255. [6.5.2]; p. 180.

Sheynin, O.B., *On the prehistory of the theory of probability*, <u>Arch. Hist. Exact Sci.</u>, 12 (1974) 97-141. [1.2.0]; p. 5.

Shinbrot, Marvin, *Fixed-point theorems*, <u>Scientific American</u>, 214 (January 1966) 105-110, 136; also in M. Kline, <u>Mathematics in the Modern World</u>, Freeman, 1968, pp. 145-150, 398. [5.5.1]; p. 140.

Shiu, P., *How slowly can a series converge?*, <u>Math. Gazette</u>, 56 (1972) 285-288. [4.2.2]; p. 97.

Shoenfield, J.R., *Martin's axiom*, <u>Amer. Math. Monthly</u>, 82 (1975) 610-617. [2.1.3]; p. 41.

Shubik, Martin, *Games of status*, <u>Behavioral Science</u>, 16 (1971) 117-129. [7.3.2]; p. 203.

Shubnikov, A.V. and Koptsik, V.A., <u>Symmetry in Science and Art</u>, Plenum Pr., 1974. [3.5.1]; p. 79.

Shul'Geifer, E.G., Kurosh, A.G. and Livshits, A. Kh., *Foundations of the theory of categories*, <u>Russian Math. Surveys</u>, 15:6 (1960) 1-46. [3.6.2]; p. 86.

Sierpiński, W., *On some unsolved problems of arithmetic*, <u>Scripta Math.</u>, 25 (1960) 125-136. [3.3.1]; p. 69.

Silver, Jack H., *The bearing of large cardinals on constructibility*, in M.D. Morley, <u>Studies in Model Theory</u>, M.A.A., 1973, pp. 158-182. [2.1.3]; p. 41.

Singh, Jagjit, <u>Great Ideas of Modern Mathematics: Their Nature and Use</u>, Dover, 1959. [1.1.0]; p. 1.

Sinkov, Abraham, <u>Elementary Cryptanalysis--A Mathematical Approach</u>, New Math. Libr., No. 22, Random House, 1968; M.A.A., 1975. [7.5.1]; p. 208.

Siu, M.K. and Lam, T.Y., *K_0 and K_1--an introduction to algebraic K-theory*, <u>Amer. Math. Monthly</u>, 82 (1975) 329-364. [3.6.3]; p. 88.

Slater, J.C., *Physics and the wave equation*, <u>Bull. Amer. Math. Soc.</u>, 52 (1946) 392-400. [7.1.2]; p. 190.

Smale, Stephen, *A survey of some recent developments in differential topology*, <u>Bull. Amer. Math. Soc.</u>, 69 (1963) 131-145. [5.6.3]; p. 147.

──────, *Differentiable dynamical systems*, <u>Bull. Amer. Math. Soc.</u>, 73 (1967) 747-817. [5.6.3]; p. 147.

──────, *What is global analysis?*, <u>Amer. Math. Monthly</u>, 76 (1969) 4-9. [5.6.2]; p. 146.

────── and Sneddon, Ian N., *Differential equations*, <u>Encyclopaedia Britannica</u>, 15th ed., 1974, Macropaedia V. 5, pp. 736-767. [4.5.2]; p. 111.

Smith, Cedric A.B., *Consistency in statistical inference and decision*, <u>J. Royal Stat. Soc.</u>, (Ser. B) 23 (1961) 1-37. [6.2.2]; p. 163.

Smith, David E., MacDuffee, C.C. and LeVeque, William J., *Arithmetic*, <u>Encyclopaedia Britannica</u>, 15th ed., 1974, Macropaedia V. 1, pp. 1171-1178. [3.6.0]; p. 84.

Smith, Thomas M., *Some perspectives on the early history of computers*, in Z.W. Pylyshyn, Perspectives on the Computer Revolution, Prentice-Hall, 1970, pp. 7-15. [6.5.0]; p. 178.

Smullyan, Raymond M., *Review of Kurt Gödel's On Formally Undecidable Propositions of Principia Mathematica and Related Systems*, Amer. Math. Monthly, 73 (1966) 319-322. [2.5.0]; p. 52.

———, *The continuum hypothesis*, in The Mathematical Sciences--A Collection of Essays for COSRIMS, M.I.T., 1969, pp. 252-260. [2.1.0]; p. 38.

———, *The continuum problem*, The Encyclopedia of Philosophy, Macmillan, 1967, V. 2, pp. 207-212. [2.1.1]; p. 39.

Sneddon, Ian N. and Smale, Stephen, *Differential equations*, Encyclopaedia Britannica, 15th ed., 1974, Macropaedia V. 5, pp. 736-767. [4.5.2]; p. 111.

Sobolev, S.L. and Ladyzenskaja, O.A., *Partial differential equations*, in A.D. Aleksandrov, et al., Mathematics--Its Content, Methods and Meaning, V. 2, M.I.T., 1963, pp. 3-55. [4.5.2]; p. 111.

Solomon, Herbert, *A survey of mathematical models in factor analysis*, in H. Solomon, Mathematical Thinking in the Measurement of Behavior, Free Pr., 1960, pp. 273-314. [7.3.1]; p. 202.

Spanier, E.H., *Grammars and languages*, Amer. Math. Monthly, 76 (1969) 335-342. [7.6.2]; p. 212.

——— and Dubins, L.E., *How to cut a cake fairly*, Amer. Math. Monthly, 68 (1961) 1-17. [6.3.3]; p. 169.

Spohn, William G., Jr., *Can mathematics be saved?*, Notices Amer. Math. Soc., 16 (1969) 890-894. [1.5.0]; p. 28.

Stauduhar, Richard P., *The determination of Galois groups*, Math. of Computation, 27 (1973) 981-996. [3.5.2]; p. 81.

Stečkin, S.B., *Theory of functions of a real variable*, in A.D. Aleksandrov, et al., Mathematics--Its Content, Methods and Meaning, V. 3, M.I.T., 1963, pp. 3-36. [4.3.1]; p. 99.

Steen, Lynn Arthur, *Conjectures and counterexamples in metrization theory*, Amer. Math. Monthly, 79 (1972) 113-132. [5.5.2]; p. 142.

———, *Foundations of mathematics: unsolvable problems*, Science, 189 (1975) 209-210. [2.5.0]; p. 52.

———, *Highlights in the history of spectral theory*, Amer. Math. Monthly, 80 (1973) 359-381. [4.6.2]; p. 117.

———, *New models of the real-number line*, Scientific American, 225 (August 1971) 92-99, 120. [2.4.1]; p. 50.

———, *Order from chaos*, Science News, 107 (1975) 292-293. [6.1.0]; p. 155.

———, *Solving the great bubble mystery*, Science News, 108 (1975) 186-187. [7.1.0]; p. 185.

―――, *The metamathematical world of model theory*, Science News, 107 (1975) 108-111. [2.4.0]; p. 49.

―――, Seebach, J. Arthur, Jr. and Seebach, Linda A., *What is a sheaf?*, Amer. Math. Monthly, 77 (1970) 681-703. [5.6.2]; p. 146.

Steenrod, Norman E., *Cohomology operations and obstructions to extending continuous functions*, Advances in Math., 8 (1972) 371-416. [5.7.3]; p. 151.

―――― and Chinn, William G., First Concepts of Topology, New Math. Libr., No. 18, Random House, 1966; M.A.A., 1975. [5.5.1]; p. 140.

Stehney, Ann K. and Millman, R.S., *The geometry of connections*, Amer. Math. Monthly, 80 (1973) 475-500. [5.3.3]; p. 135.

Stein, Sherman K., *Algebraic tiling*, Amer. Math. Monthly, 81 (1974) 445-462. [3.5.2]; p. 81.

―――, *The mathematician as an explorer*, Scientific American, 204 (May 1961) 148-158, 206. [1.5.0]; p. 28.

Steinhaus, Hugo, Mathematical Snapshots, Oxford U. Pr., 1969. [1.1.0]; p. 1.

―――, *Stefan Banach, 1892-1945*, Scripta Math., 26 (1963) 93-100. [1.3.0]; p. 19.

Stevens, Peter S., Patterns in Nature, Atlantic-Little, Brown, 1974. [5.1.0]; p. 126.

Stewart, Ian, Concepts of Modern Mathematics, Penguin, 1975. [1.1.0]; p. 1.

―――, *The seven elementary catastrophes*, New Scientist (29 November 1975) 447-454. [5.6.1]; p. 145.

Stoll, Robert S. and Hashisaki, Joseph, *Set theory*, Encyclopaedia Britannica, 15th ed., 1974, Macropaedia V. 16, pp. 569-575. [2.1.1]; p. 39.

Stolzenberg, Gabriel, *Review of Errett Bishop's Foundations of Constructive Analysis*, Bull. Amer. Math. Soc., 76 (1970) 301-323. [2.3.1]; p. 48.

Stone, Marshall H., *Mathematics and the future of science*, Bull. Amer. Math. Soc., 63 (1957) 61-76. [1.4.0]; p. 22.

―――, *The future of mathematics*, J. Math. Soc. Jap., 9 (1957) 493-507. [1.5.0]; p. 28.

―――, *The revolution in mathematics*, Liberal Education, 47 (1961) 304-327; also in Amer. Math. Monthly, 68 (1961) 715-734. [1.1.0]; p. 1.

Strang, Gilbert, *The finite element method--linear and nonlinear applications*, in Proc. Inter. Cong. Math. (1974), V. 2, Canad. Cong. Math., 1975, pp. 429-435. [6.4.2]; p. 175.

Struik, Dirk J., *Conic sections*, Encyclopedia Americana, 1976, V. 7, pp. 578-579. [5.3.1]; p. 133.

――――, *Descriptive geometry*, Encyclopedia Americana, 1976, V. 12, pp. 492-494. [5.1.1]; p. 127.

――――, *Differential geometry*, Encyclopedia Americana, 1976, V. 12, pp. 499-501. [5.3.1]; p. 133.

――――, *On the foundations of the theory of probabilities*, Phil. Sci., 1 (1934) 50-70. [6.1.1]; p. 156.

Struik, Ruth R. and Fisher, Irwin, *Nil algebras and periodic groups*, Amer. Math. Monthly, 75 (1968) 611-623. [3.5.3]; p. 82.

Sturgis, H.E. and Shepherdson, J.C., *Computability of recursive functions*, J. Assoc. Comp. Mach., 10 (1963) 217-255. [6.5.2]; p. 180.

Sullivan, Walter, *New form of math said to resolve old paradoxes of numbering system*, New York Times (February 15, 1975) 22. [2.4.0]; p. 49.

Suppes, P., *A comparison of the meaning and uses of models in mathematics and the empirical sciences*, Synthese, 12 (1960) 287-301. [1.5.0]; p. 28.

Szabó, Árpád, *The transformation of mathematics into a deductive science and the beginnings of its foundations on definitions and axioms*, Scripta Math., 27 (1964) 28-48A, 113-139. [1.5.0]; p. 28.

Tahta, D.G., *A startling discovery*, Math. Teaching, 45 (1968) 33-37. [4.2.1]; p. 95.

Tanur, Judith M., et al., *Statistics: A Guide to the Unknown*, Holden-Day, 1972. [6.2.0]; p. 161.

Tarski, Alfred, *Truth and proof*, Scientific American, 220 (June 1969) 63-77, 144. [2.6.0]; p. 55.

―――― and Keisler, H. Jerome, *From accessible to inaccessible cardinals*, Fund. Math., 53 (1963) 225-308. [2.1.3]; p. 40.

Tate, J., *The 1974 Fields Medals (I): An algebraic geometer*, Science, 186 (1974) 39-40. [1.3.2]; p. 20.

Taub, A.H., *Relativistic hydrodynamics*, in A.H. Taub, Studies in Applied Mathematics, M.A.A., 1971, pp. 150-180. [7.1.3]; p. 191.

Taussky, Olga, *Sums of squares*, Amer. Math. Monthly, 77 (1970) 805-830. [3.3.2]; p. 71.

Taylor, Angus E., *Differential and integral calculus*, McGraw-Hill Encyclopedia of Science and Technology, 1960, V. 2, pp. 401-403. [4.1.1]; p. 92.

――――, *Differentiation*, McGraw-Hill Encyclopedia of Science and Technology, 1960, V. 4, pp. 128-132. [4.1.1]; p. 92.

――――, *Notes on the history of the uses of analyticity in operator theory*, Amer. Math. Monthly, 78 (1971) 331-342. [4.6.3]; p. 119.

――――, *Some aspects of mathematical research*, Amer. Scientist, 35 (1947) 211-223. [1.5.1]; p. 30.

———, *The differential: nineteenth and twentieth century developments*, Arch. Hist. Exact Sci., 12 (1974) 355-383. [1.2.1]; p. 9.

Taylor, H.M. and Costello, W.G., *Deterministic population growth models*, Amer. Math. Monthly, 78 (1971) 841-855. [7.2.2]; p. 198.

Taylor, Joseph L., *Measure algebras*, CBMS Reg. Conf. Ser. in Math., No. 16, Amer. Math. Soc., 1972. [4.7.3]; p. 124.

Temple, G., *The growth of mathematics*, Math. Gazette, 41 (1957) 161-168. [1.1.0]; p. 1.

———, *Theories and applications of generalized functions*, J. London Math. Soc., 28 (1953) 134-148. [4.6.2]; p. 117.

Tewarson, R.P., *Computations with sparse matrices*, SIAM Review, 12 (1970) 527-543. [6.4.2]; p. 175.

The mathemagician, Time (April 21, 1975) 63. [1.3.0]; p. 19.

The state of the mathematical sciences, in The Mathematical Sciences --A Report, National Academy of Sciences, 1968, pp. 45-116. [1.1.0]; p. 2.

Thébault, Victor, *Geodesics*, Scripta Math., 21 (1955) 146-158. [5.3.1]; p. 133.

Theory of probability, Encyclopaedia Britannica, 15th ed., 1974, Macropaedia V. 14, pp. 1104-1115. [6.1.1]; p. 156.

Thom, René F., *"Modern" mathematics: an educational and philosophic error?*, Amer. Scientist, 59 (1971) 695-699. [1.6.0]; p. 34.

———, *Topological models in biology*, Topology, 8 (1969) 313-335. [7.2.2]; p. 199.

——— and Mostow, George D., *Topological groups and differential topology*, Encyclopaedia Britannica, 15th ed., 1974, Macropaedia V. 18, pp. 489-504. [5.8.3]; p. 152.

Thomas, T.Y. and Mayer, W., *Foundations of the theory of Lie groups*, Annals of Math., 36 (1935) 770-822. [5.8.3]; p. 152.

Thompson, Gerald L., *Game theory*, McGraw-Hill Encyclopedia of Science and Technology, 1960, V. 6, pp. 24-30. [6.3.1]; p. 167.

Threlfall, William and Seifert, Herbert, *Old and new results on knots*, Canad. J. Math., 2 (1950) 1-15. [5.7.3]; p. 150.

Thurston, Hugh, *Series*, Encyclopedia Americana, 1976, V. 24, pp. 576-580. [4.2.1]; p. 95.

———, *Tangents: an elementary survey*, Math. Magazine, 42 (1969) 1-11. [5.1.1]; p. 128.

———, *What exactly is dy/dx?*, Educ. Studies Math., 4 (1972) 358-367. [4.1.1]; p. 92.

Tietze, Heinrich, Famous Problems of Mathematics, Graylock Pr., 1965. [1.1.1]; p. 2.

Toeplitz, Otto and Rademacher, Hans, The Enjoyment of Mathematics: Selections from Mathematics for the Amateur, Princeton U. Pr., 1957. [1.1.1]; p. 2.

Tolsted, Elmer, *An elementary derivation of the Cauchy, Hölder, and Minkowski inequalities from Young's inequality*, Math. Magazine, 37 (1964) 2-12. [4.3.2]; p. 101.

Tompkins, C.B., *Calculus of variations*, McGraw-Hill Encyclopedia of Science and Technology, 1960, V. 2, pp. 407-410. [6.3.1]; p. 167.

Tóth, Imre, *Non-Euclidean geometry before Euclid*, Scientific American, 221 (November 1969) 87-98, 166. [5.1.0]; p. 126.

Trakhtenbrot, B.A., *Algorithms*, in Z.W. Pylyshyn, Perspectives on the Computer Revolution, Prentice-Hall, 1970, pp. 69-86. [6.5.1]; p. 179.

Treves, François, *Applications of distributions to PDE theory*, Amer. Math. Monthly, 77 (1970) 241-248. [4.5.2]; p. 111.

———, *On local solvability of linear partial differential equations*, Bull. Amer. Math. Soc., 76 (1970) 552-571. [4.5.3]; p. 112.

———, Schwartz, Jacob T. and Morrey, Charles B., Jr., *Functional analysis*, Encyclopaedia Britannica, 15th ed., 1974, Macropaedia V. 1, pp. 757-772. [4.6.2]; p. 117.

Tucker, Alan, *Pólya's enumeration formula by example*, Math. Magazine, 47 (1974) 248-256. [3.1.2]; p. 61.

Tucker, Albert W., *Combinatorial algebra of linear programs*, in J.G. Kemeny, R. Robinson and R.W. Ritchie, New Directions in Mathematics, Prentice-Hall, 1961, pp. 77-91; also in G.B. Dantzig and B.C. Eaves, Studies in Optimization, M.A.A., 1974, pp. 9-26. [6.3.1]; p. 167.

———, *Some topological properties of disk and sphere*, in Proc. First Canad. Math. Cong., U. Toronto Pr., 1946, pp. 285-309. [5.5.2]; p. 142.

———, et al., *Mathematical theory of optimization*, Encyclopaedia Britannica, 15th ed., 1974, Macropaedia V. 13, pp. 621-638. [6.3.1]; p. 167.

——— and Bailey, Herbert S., Jr., *Topology*, Scientific American, 182 (January 1950) 18-24, 64; also in M. Kline, Mathematics in the Modern World, Freeman, 1968, pp. 134-140, 398. [5.5.0]; p. 139.

——— and Kuhn, Harold W., *John von Neumann's work in the theory of games and mathematical economics*, Bull. Amer. Math. Soc., 64 (1958) Suppl. pp. 100-122. [1.3.1]; p. 19.

Tukey, John W., *Mathematics and the picturing of data*, in Proc. Inter. Cong. Math. (1974), V. 2, Canad. Cong. Math., 1975, pp. 523-531. [6.2.1]; p. 162.

Turner, James and Kautz, William H., *A survey of progress in graph theory in the Soviet Union*, SIAM Review, 12 (1970) Suppl. pp. 1-68. [3.2.2]; p. 64.

Tutte, W.T., *Chromials*, in D.R. Fulkerson, Studies in Graph Theory, Part II, M.A.A., 1975, pp. 361-377. [3.2.2]; p. 64.

———, *Map-coloring problems and chromatic polynomials*, Amer. Scientist, 62 (1974) 702-705. [3.2.1]; p. 64.

———, *Symmetrical graphs and coloring problems*, Scripta Math., 25 (1960) 305-316. [3.2.1]; p. 64.

——— and Whitney, Hassler, *Kempe chains and the four color problem*, Utilitas Mathematica, 2 (November 1972); also in D.R. Fulkerson, Studies in Graph Theory, Part II, M.A.A., 1975, pp. 378-413. [3.2.2]; p. 65.

Udell, Dan E., *Self-similar models for erratic chance processes*, IBM Research Reports, 2:2 (1966) 1-4. [6.1.0]; p. 155.

Ulam, Stanislaw M., *John von Neumann, 1903-1957*, Bull. Amer. Math. Soc., 64 (1958) Suppl. pp. 1-49. [1.3.2]; p. 20.

———, *Monte Carlo calculations in problems of mathematical physics*, in E.F. Beckenbach, Modern Mathematics for the Engineer, 2nd Ser., McGraw-Hill, 1961, pp. 261-281. [6.1.2]; p. 159.

———, *What is measure?*, Amer. Math. Monthly, 50 (1943) 597-602. [4.3.1]; p. 99.

——— and Kac, Mark, Mathematics and Logic: Retrospect and Prospects, Frederick A. Praeger, 1968; The New American Library, 1969. [1.1.2]; p. 3.

Ullian, Joseph S., *Is any set theory true?*, Phil. Sci., 36 (1969) 271-279. [2.2.2]; p. 46.

Valentine, Frederick A., *Visible shorelines*, Amer. Math. Monthly, 77 (1970) 146-152. [5.2.2]; p. 131.

——— and Forder, Henry G., *Euclidean geometry*, Encyclopaedia Britannica, 15th ed., 1974, Macropaedia V. 7, pp. 1099-1112. [5.1.1]; p. 127.

van der Waerden, B.L., *The foundation of algebraic geometry from Severi to André Weil*, Arch. Hist. Exact Sci., 7 (1971) 171-180. [1.2.3]; p. 13.

Vandiver, H.S., *Fermat's last theorem--its history and the nature of the known results concerning it*, Amer. Math. Monthly, 53 (1946) 555-578. [3.3.2]; p. 72.

Van Osdol, D.H., *Truth with respect to an ultrafilter or how to make intuition rigorous*, Amer. Math. Monthly, 79 (1972) 355-363. [2.4.2]; p. 50.

Varga, Richard S., *Iterative methods for solving matrix equations*, Amer. Math. Monthly, 72 (1965) Suppl. pp. 67-74. [6.4.2]; p. 175.

Varma, R.S. and Nath, Prem, *Information theory--a survey*, J. Math. Sci., 2 (1967) 75-109. [7.5.2]; p. 209.

Vaught, Robert L., *Models of complete theories*, Bull. Amer. Math. Soc., 69 (1963) 299-313. [2.6.2]; p. 57.

———, *Some aspects of the theory of models*, Amer. Math. Monthly, 80 (1973) Suppl. pp. 3-37. [2.6.3]; p. 58.

Vilenkin, N. Ya, Stories about Sets, Academic Pr., 1968. [2.1.0]; p. 38.

Viscensini, P., *Differential geometry in the nineteenth century*, Scientia, 107 (1972) 661-696. [5.3.3]; p. 135.

Vitány, Paul M.B. and Herman, Gabor T., *Growth functions associated with biological development*, Amer. Math. Monthly, 83 (1976) 1-15. [7.2.2]; p. 198.

von Mises, Richard, Probability, Statistics and Truth, Macmillan, 1957. [6.1.0]; p. 155.

von Neumann, John, *The mathematician*, in R.B. Heywood, The Works of the Mind, U. of Chicago Pr., 1947, pp. 180-196; also in J.R. Newman, The World of Mathematics, V. 4, Simon and Schuster, 1956, pp. 2053-2063. [1.5.0]; p. 28.

Vorob'ev, N.N., *The present state of the theory of games*, Russian Math. Surveys, 25:2 (1970) 77-136. [6.3.2]; p. 169.

Waismann, Friedrich, Introduction to Mathematical Thinking, Harper & Brothers, 1959. [1.1.0]; p. 2.

Walsh, Bertram, *The scarcity of cross products on Euclidean spaces*, Amer. Math. Monthly, 74 (1967) 188-194. [3.4.1]; p. 76.

Walsh, J.L., *Complex numbers and complex variables*, McGraw-Hill Encyclopedia of Science and Technology, 1960, V. 3, pp. 336-343. [4.4.1]; p. 104.

———, *Conformal mapping*, McGraw-Hill Encyclopedia of Science and Technology, 1960, V. 3, pp. 395-397. [4.4.1]; p. 104.

———, *History of the Riemann mapping theorem*, Amer. Math. Monthly, 80 (1973) 270-276. [4.4.3]; p. 107.

Waterman, Daniel and Goffman, Casper, *Some aspects of Fourier series*, Amer. Math. Monthly, 77 (1970) 119-133. [4.3.3]; p. 102.

Waugh, Frederick V. and Maxfield, Margaret W., *Side-and-diagonal numbers*, Math. Magazine, 40 (1967) 74-83. [3.3.1]; p. 69.

Weaver, Warren, *Probability*, Scientific American, 183 (October 1950) 44-47, 64; also in M. Kline, Mathematics in the Modern World, Freeman, 1968, pp. 161-164, 398. [6.1.0]; p. 155.

———, *Statistics*, Scientific American, 186 (January 1952) 60-63, 84; also in M. Kline, Mathematics in the Modern World, Freeman, 1968, pp. 175-178, 399. [6.2.0]; p. 162.

Weidman, Donald R., *Emotional perils of mathematics*, Science, 149 (1965) 1048. [1.5.0]; p. 28.

Weil, André, *Algebraic geometry*, Encyclopedia Americana, 1976, V. 12, pp. 501-503. [5.4.2]; p. 137.

———, *The future of mathematics*, Amer. Math. Monthly, 57 (1950) 295-306; also in F. LeLionnais, Great Currents of Mathematical Thought, V. 1, Dover, 1971, pp. 321-336. [1.1.1]; p. 2.

———, *Two lectures on number theory, past and present*, L'Enseignement Math., 20 (1974) 87-110. [3.3.1]; p. 69.

Weiss, Guido, *Complex methods in harmonic analysis*, Amer. Math. Monthly, 77 (1970) 465-474. [4.7.3]; p. 124.

———, *Harmonic analysis*, in I.I. Hirschman, Jr., Studies in Real and Complex Analysis, M.A.A., 1965, pp. 124-178. [4.7.2]; p. 122.

——— and Coifman, R.R., *Representations of compact groups and spherical harmonics*, L'Enseignement Math., 14 (1968) 121-173. [4.7.3]; p. 122.

Weissinger, Johannes, *The characteristic features of mathematical thought*, in T.L. Saaty and F.J. Weyl, The Spirit and Uses of the Mathematical Sciences, McGraw-Hill, 1969, pp. 9-27. [1.5.0]; p. 29.

Wermer, John, *Banach algebras and analytic functions*, Advances in Math., 1 (1965) 51-102. [4.6.3]; p. 119.

———, *Uniform approximation and maximal ideal spaces*, Bull. Amer. Math. Soc., 68 (1962) 298-305. [4.6.3]; p. 119.

West, Jerry L., *The Cantor set*, Pi Mu Epsilon Journal, 5 (1970) 119-123. [4.2.2]; p. 97.

Weyl, Hermann, *A half-century of mathematics*, Amer. Math. Monthly, 58 (1951) 523-553. [1.1.1]; p. 2.

———, *Emmy Noether*, Scripta Math., 3 (1935) 201-220. [1.3.0]; p. 19.

———, *Insight and reflection*, Studia Philosophica, 15 (1955); also in T.L. Saaty and F.J. Weyl, The Spirit and Uses of the Mathematical Sciences, McGraw-Hill, 1969, pp. 281-301. [1.5.0]; p. 29.

———, *Mathematics and logic*, Amer. Math. Monthly, 53 (1946) 2-13. [2.6.1]; p. 56.

———, *Ramifications, old and new, of the eigenvalue problem*, Bull. Amer. Math. Soc., 56 (1950) 115-139. [4.5.2]; p. 111.

———, *Relativity theory as a stimulus in mathematical research*, Proc. Amer. Phil. Soc., 93 (1949) 535-541. [1.2.2]; p. 12.

———, Symmetry, Princeton U. Pr., 1952; excerpted in J.R. Newman, The World of Mathematics, V. 1, Simon and Schuster, 1956, pp. 671-724. [1.4.1]; p. 23.

———, *The mathematical way of thinking*, Science, 92 (1940) 437-446; also in Studies in the History of Science, U. Penn. Pr., 1941, pp. 103-123. [1.5.0]; p. 29.

White, William W., *A status report on computing algorithms for mathematical programming*, Computing Surveys, 5 (1973) 135-166. [6.5.3]; p. 181.

Whitehead, Alfred North, An Introduction to Mathematics, Oxford U. Pr., 1958. [1.1.0]; p. 2.

———, *Mathematics and liberal education*, in A.N. Whitehead, Essays in Science and Philosophy, Philosophical Library, 1947, pp. 175-188. [1.4.0]; p. 22.

――――, *Mathematics as an element in the history of thought*, in J.R. Newman, The World of Mathematics, V. 1, Simon and Schuster, 1956, pp. 402-416. [1.5.0]; p. 29.

Whitham, G.B., *Dispersive waves and variational principles*, in A.H. Taub, Studies in Applied Mathematics, M.A.A., 1971, pp. 181-212. [7.1.3]; p. 191.

Whitney, Hassler, *The mathematics of physical quantities*, Amer. Math. Monthly, 75 (1968) 115-138, 227-256. [7.1.2]; p. 190.

―――― and Tutte, W.T., *Kempe chains and the four color problem*, Utilitas Mathematica, 2 (November 1972); also in D.R. Fulkerson, Studies in Graph Theory, Part II, M.A.A., 1975, pp. 378-413. [3.2.2]; p. 65.

Whyburn, Gordon T., *Developments in topological analysis*, Fund. Math., 50 (1962) 305-318. [4.4.2]; p. 106.

――――, *On the structure of continua*, Bull. Amer. Math. Soc., 42 (1936) 49-73. [5.5.3]; p. 143.

――――, *Topological analysis*, Bull. Amer. Math. Soc., 62 (1956) 204-218. [4.3.2]; p. 101.

――――, *What is a curve?*, Amer. Math. Monthly, 49 (1942) 493-497; also in P.J. Hilton, Studies in Modern Topology, M.A.A., 1968, pp. 23-38. [5.5.2]; p. 142.

Widder, D.V. and Hirschman, I.I., Jr., *The Laplace transform, the Stieltjes transform, and their generalizations*, in I.I. Hirschman, Jr., Studies in Real and Complex Analysis, M.A.A., 1965, pp. 67-89. [4.2.2]; p. 97.

Wiener, Norbert, Ex-Prodigy, Simon and Schuster, 1953; M.I.T. Pr., 1964. [1.3.0]; p. 19.

――――, I am a Mathematician, Doubleday, 1956; M.I.T. Pr., 1964. [1.3.0]; p. 19.

――――, *The theory of prediction*, in E.F. Beckenbach, Modern Mathematics for the Engineer, McGraw-Hill, 1956, pp. 165-190. [4.7.3]; p. 124.

Wightman, A.S., *Analytic functions and elementary particles*, in The Mathematical Sciences--A Collection of Essays for COSRIMS, M.I.T., 1969, pp. 116-127. [4.4.1]; p. 104.

――――, *Group theory*, McGraw-Hill Encyclopedia of Science and Technology, 1960, V. 6, pp. 282-285. [3.5.1]; p. 80.

Wigner, Eugene P., *Symmetry principles in old and new physics*, Bull. Amer. Math. Soc., 74 (1968) 793-815. [7.1.2]; p. 190.

――――, *The unreasonable effectiveness of mathematics in the natural sciences*, Comm. Pure Appl. Math., 13 (1960); also in T.L. Saaty and F.J. Weyl, The Spirit and Uses of the Mathematical Sciences, McGraw-Hill, 1969, pp. 123-140; in Studies in Mathematics, V. 16, SMSG, 1967, pp. 31-44; and in E.P. Wigner, Symmetries and Reflections: Scientific Essays of Eugene P. Wigner, Indiana U. Pr., 1967, pp. 222-237. [1.4.0]; p. 22.

Wilansky, Albert, *Spectral decomposition of matrices for high school students*, Math. Magazine, 41 (1968) 51-59. [3.4.1]; p. 76.

Wilder, Raymond L., Evolution of Mathematical Concepts: An Elementary Study, John Wiley, 1968. [1.2.0]; p. 5.

———, *Hereditary stress as a cultural force in mathematics*, Historia Math., 1 (1974) 29-46. [1.2.0]; p. 6.

———, *History in the mathematics curriculum: its status, quality, and function*, Amer. Math. Monthly, 79 (1972) 479-495. [1.6.1]; p. 35.

———, *The nature of mathematical proof*, Amer. Math. Monthly, 51 (1944) 309-323. [1.5.1]; p. 30.

———, *The origin and growth of mathematical concepts*, Bull. Amer. Math. Soc., 59 (1963) 423-448. [1.5.1]; p. 30.

———, *The role of the axiomatic method*, Amer. Math. Monthly, 74 (1967) 115-127; also in Math. Teaching, 41 (1967) 32-40. [1.5.0]; p. 29.

———, *Topology--its nature and significance*, Math. Teacher, 55 (1962) 462-475. [5.5.1]; p. 140.

———, *Trends and social implications of research*, Bull. Amer. Math. Soc., 75 (1969) 891-906. [1.4.0]; p. 22.

Wilkinson, J.H., *Modern error analysis*, SIAM Review, 13 (1971) 548-568. [6.4.2]; p. 175.

Williamson, J.H., *Harmonic analysis on semigroups*, J. London Math. Soc., 42 (1967) 1-41. [4.7.3]; p. 124.

Willoughby, Stephen S., *Analytic geometry*, Encyclopedia Americana, 1976, V. 12, pp. 483-487. [5.1.1]; p. 128.

———, *Euclidean geometry*, Encyclopedia Americana, 1976, V. 12, pp. 479-483. [5.1.1]; p. 128.

Wilson, Curtis, *How did Kepler discover his first two laws?*, Scientific American, 226 (March 1972) 92-106, 126. [1.2.0]; p. 6.

Wilson, R.J., *An introduction to matroid theory*, Amer. Math. Monthly, 80 (1973) 500-525. [3.2.2]; p. 65.

Wyler, Oswald, *Exterior differential calculus and Maxwell's equations*, in K.O. May, Lectures on Calculus, Holden-Day, 1967, pp. 147-165. [4.2.2]; p. 98.

Wylie, C. Ray, *Surface*, Encyclopedia Americana, 1976, V. 26, pp. 50-51. [5.3.1]; p. 133.

Wyman, B.F., *What is a reciprocity law?*, Amer. Math. Monthly, 79 (1972) 571-586; Correction, 80 (1973) 281. [3.3.3]; p. 73.

Yaglom, I.M., Geometric Transformations, New Math. Libr., No. 8, Random House, 1962; M.A.A., 1975. [5.1.1]; p. 128.

———, Geometric Transformations II, New Math. Libr., No. 21, Random House, 1968; M.A.A., 1975. [5.1.1]; p. 128.

———, *Geometric Transformations III*, New Math. Libr., No. 24, Random House, 1973; M.A.A., 1975. [5.1.1]; p. 128.

Yale, Paul B., *Automorphisms of the complex numbers*, Math. Magazine, 39 (1966) 135-141. [3.6.2]; p. 86.

Youden, William J., *How mathematics appraises risks and gambles*, in T.L. Saaty and F.J. Weyl, The Spirit and Uses of the Mathematical Sciences, McGraw-Hill, 1969, pp. 167-187. [6.1.0]; p. 155.

Young, David M., *A survey of modern numerical analysis*, SIAM Review, 15 (1973) 503-523. [6.4.2]; p. 175.

Young, H.P. and Balinski, M.L., *The quota method of apportionment*, Amer. Math. Monthly, 82 (1975) 701-730. [7.3.1]; p. 201.

Zalcman, Lawrence, *Real proofs of complex theorems (and vice versa)*, Amer. Math. Monthly, 81 (1974) 115-137. [4.4.2]; p. 106.

Zassenhaus, Hans J., *Modern developments in the geometry of numbers*, Bull. Amer. Math. Soc., 67 (1961) 427-439. [5.2.3]; p. 132.

———, *On the fundamental theorem of algebra*, Amer. Math. Monthly, 74 (1967) 485-497. [3.6.2]; p. 86.

———, *What is an angle?*, Amer. Math. Monthly, 61 (1954) 369-378. [5.1.0]; p. 126.

Zeeman, E. Christopher, *Catastrophe theory*, Scientific American, 234 (April 1976) 65-83, 138. [7.2.1]; p. 197.

———, *Levels of structure in catastrophe theory illustrated by applications in the social and biological sciences*, in Proc. Inter. Cong. Math. (1974), V. 2, Canad. Cong. Math., 1975, pp. 533-546. [7.3.2]; p. 203.

———, *The geometry of catastrophe*, Times Lit. Suppl., (10 December 1971) 1556-1557. [5.6.1]; p. 145.

Zippin, Leo, Uses of Infinity, New Math. Libr., No. 7, Random House, 1962; M.A.A., 1975. [2.1.0]; p. 38.

Zlot, William Leonard, *The principle of choice in preaxiomatic set theory*, Scripta Math., 25 (1960) 105-123. [2.1.2]; p. 40.

Zobrist, Albert L. and Carlson, Frederic R., Jr., *An advice-taking chess computer*, Scientific American, 228 (June 1973) 92-105, 124. [6.5.0]; p. 178.

Zuckerman, H.S. and Niven, Ivan M., *Lattice points and polygonal area*, Amer. Math. Monthly, 74 (1967) 1195-1200. [5.1.1]; p. 127.

Zygmund, Antoni and Fefferman, Charles L., *Fourier analysis*, Encyclopaedia Britannica, 15th ed., 1974, Macropaedia V. 1, pp. 735-757. [4.7.2]; p. 122.

Ref
Z
6651
G33

NOV 17 1976